Tainted Earth

Critical Issues in Health and Medicine

Edited by Rima D. Apple, University of Wisconsin–Madison,
and Janet Golden, Rutgers University, Camden

Growing criticism of the U.S. health care system is coming from consumers, politicians, the media, activists, and health care professionals. Critical Issues in Health and Medicine is a collection of books that explores these contemporary dilemmas from a variety of perspectives, among them political, legal, historical, sociological, and comparative, and with attention to crucial dimensions such as race, gender, ethnicity, sexuality, and culture.

For a list of titles in the series, see the last page of the book.

Tainted Earth

Smelters, Public Health, and the Environment

MARIANNE SULLIVAN

RUTGERS UNIVERSITY PRESS
NEW BRUNSWICK, NEW JERSEY, AND LONDON

Library of Congress Cataloging-in-Publication Data
Sullivan, Marianne, 1970–
 Tainted earth : smelters, public health, and the environment / Marianne Sullivan.
  p. ; cm. — (Critical issues in health and medicine)
 Includes bibliographical references and index.
 ISBN 978–0–8135–6279–7 (hardcover : alk. paper) — ISBN 978–0–8135–6278–0 (pbk. : alk. paper) — ISBN 978–0–8135–6280–3 (e-book)
 I. I Title. I. Series: Critical issues in health and medicine.
 [DNLM: 1. Environmental Pollution—adverse effects—Idaho. 2. Environmental Pollution—adverse effects—Texas. 3. Environmental Pollution—adverse effects—Washington. 4. Environmental Pollution—history—Idaho. 5. Environmental Pollution—history—Texas. 6. Environmental Pollution—history—Washington. 7. Carcinogens, Environmental—toxicity—Idaho. 8. Carcinogens, Environmental—toxicity—Texas. 9. Carcinogens, Environmental—toxicity—Washington. 10. Extraction and Processing Industry—Idaho. 11. Extraction and Processing Industry—Texas. 12. Extraction and Processing Industry—Washington. 13. History, 20th Century—Idaho. 14. History, 20th Century—Texas. 15. History, 20th Century—Washington. 16. Industrial Waste—adverse effects—Idaho. 17. Industrial Waste—adverse effects—Texas. 18. Industrial Waste—adverse effects—Washington. WA 670]
 RA576.A1
 363.739'2—dc23

2013013432

A British Cataloging-in-Publication record for this book
is available from the British Library.

Copyright © 2014 by Marianne Sullivan
All rights reserved
No part of this book may be reproduced or utilized in any form or by any means, electronic or mechanical, or by any information storage and retrieval system, without written permission from the publisher. Please contact Rutgers University Press, 106 Somerset Street, New Brunswick, NJ 08901. The only exception to this prohibition is "fair use" as defined by U.S. copyright law.

Visit our website: http://rutgerspress.rutgers.edu

Manufactured in the United States of America

For James

CONTENTS

List of Figures   ix
Acknowledgments   xi

	Introduction	1
1	The Tacoma Smelter	11
2	City of Destiny, City of Smoke	31
3	Uncovering a Crisis in El Paso	55
4	Bunker Hill	73
5	Tacoma: A Disaster Is Discovered	111
6	A Carcinogenic Threat	129
7	Sacrificed	155
	Conclusion	171

Notes   175
Index   227

FIGURES

Figure 1.1	Puget Sound Region of Washington State	12
Figure 2.1	Tacoma smelter and surrounding community, 1953	49
Figure 3.1	ASARCO El Paso site and surrounding communities	57
Figure 3.2	CDC blood lead testing in El Paso, 1972	63
Figure 4.1	Pollution pours out of the Bunker Hill smelter, 1969–1970	73
Figure 4.2	Coeur d'Alene mining district	78
Figure 4.3	The 3.3 millionth ton of lead	79
Figure 5.1	Dr. Samuel Milham in Ruston, 1972	122
Figure 7.1	Arsenic soil contamination in the Puget Sound Region	161

ACKNOWLEDGMENTS

There are many people who helped me on the path to completing this book. Foremost among them is my family. My husband, James, took our boys on numerous day trips and weekend-long outings to provide me with the quiet, child-free space necessary for writing. The fact that this project is now complete has much to do with those trips and outings, combined with gentle exhortations to "get it done." My parents helped immensely by providing child care and moral support, and my English teacher mother provided some greatly appreciated editorial assistance.

The idea for this book started in David Rosner's "History of Public Health" seminar at the Mailman School of Public Health at Columbia University. His early support and intellectual guidance contributed significantly to the finished product. Gerald Markowitz has given freely of his time, providing constructive criticism, advice, and encouragement. Other faculty members at the Mailman School of Public Health also helped me to think through key problems.

Just as smelters and their pollution shaped landscapes and communities, they also fundamentally shaped many of those who studied their environmental and public health toll—their convictions, careers, worldviews, and dedication to their professions in public health, medicine, environmental science and law, and environmental activism. This quickly became apparent to me when I contacted many of the people who were deeply engaged in these crises in the 1970s and 1980s and found that they could not have been more willing to share their knowledge, expertise, and experiences.

Judy Alsos supported my work from the beginning and invited me into her North Tacoma home on numerous occasions to share her recollections of community organizing in the 1950s and 1960s in Tacoma. When I emailed Paul Whelan to ask some specific questions about Bunker Hill, even though he did not know me in the least, my phone rang not five minutes later and he talked openly with me about his experiences helping injured children get compensation. He also welcomed me into his Seattle law office to share his nonconfidential files from the case. Even though I did not meet them in person, I am grateful to the other members of the Yoss legal team, in particular Susan Lee and Larry Axtell, for the thoroughness of their background research while developing the case. Their dogged research and questioning in the late 1970s helped to establish the record of what happened at Bunker Hill. They were decades ahead of the public health community in confronting the role of industry, trade groups, and public relations firms working together to shape the public discourse about risk from smelters.

Greg Glass generously shared with me his encyclopedic knowledge of the Tacoma smelter and its pollution problems. Washington State residents owe him much thanks for helping to meticulously document the smelter's soil contamination footprint, which led to a significant settlement for Washington State from the American Smelting and Refining Company (ASARCO) bankruptcy.

Lin Nelson and Anne Fischel have been great friends and colleagues who also have been working on public health issues in smelting communities. We have shared materials and stories, and they have kept me connected to current issues in smelting communities, particularly worker health issues.

Thank you to the numerous archivists and staff members at the National Archives and Records Administration in Seattle, the Washington State archives, the King County archives, the University of Washington Library's Special Collections, the University of Idaho, EPA Region 10 Superfund records, and the Tacoma Public Library who have helped me to access documents and photographs.

Rutgers University Press has been extremely helpful in this book coming to fruition. Doreen Valentine initially gave critical feedback, and Peter Mickulas has helped immeasurably to shape and hone the final version. The reviewers provided significant critical feedback and helped to strengthen the manuscript considerably.

Colleagues at Hofstra University and William Paterson University have been uniformly supportive of my preoccupation with completing this project.

Small portions of chapters 5 and 6 were originally published as "Contested Science and Exposed Workers," *Public Health Reports* 122 (July–August 2007), 541–547. Thank you to the editors of *Public Health Reports* for allowing republication here.

Research on the Tacoma Smelter was supported by the United States Environmental Protection Agency's Science to Achieve Results (STAR) Graduate Fellowship Program.

Tainted Earth

Introduction

On a rare sunny January day in Ruston, Washington, hundreds of people lined the town's streets and hillsides to catch a glimpse of destruction. Two miles away, across Puget Sound, on the south end of Vashon Island, crowds also stood waiting, binoculars pressed to their eyes, for the same reason. In between the mainland and the island, others surveyed the Ruston shoreline from their boats anchored in Commencement Bay. An estimated seventy thousand people turned out to watch, and many of the gatherings had a celebratory air. At the appointed time, a twelve-year-old boy pushed a plunger, demolition experts ignited dynamite charges, and the American Smelting and Refining Company (ASARCO) smelter's massive 562-foot stack toppled to the ground in seconds. Oblivious to the mixed emotions of his elders, the twelve-year-old representative of a new era said, "I'm not going to miss it. It polluted the air and stuff."[1]

The smokestack, once admired for its architecture and impressive size, a monument to U.S. twentieth-century industrial power, was reduced to rubble, covered in a huge cloud of arsenic-laden dust. As the stack fell, onlookers cheered. Was it because it was awe-inspiring to see such a massive manmade edifice reduced to rubble in just a few seconds? Or were they cheering because the stack would never again be able to rain down sulfur dioxide, sulfuric acid, arsenic, lead, mercury, and cadmium on their homes and yards? Or was it because the stack demolition seemed to signal an end to the old fights over the smelter and its pollution, pointing the way toward reinvention and renewal—an escape from a choking industrial past?

The date was January 17, 1993. The smokestack, in some iteration, had towered over North Tacoma and the tiny town of Ruston for about a century, spreading its pollution at least as far as Seattle, about thirty miles to the north. While operating, the smelter employed many of Ruston's residents and bolstered the region's tax base. But in the space of a century, the smokestack went from being nearly an unalloyed good, a symbol of jobs and economic prosperity, to being considered by many a serious threat to public health and the environment. How did an industry, whose considerable pollution was officially considered nothing more than a nuisance for many decades, come to be regarded as a threat to the health, development, and life chances of people living in the shadow of its stacks?

Smelter smokestacks once dotted the western landscape in states like Idaho, Texas, Arizona, Utah, and California, taking in raw ore and turning out pure metal—the copper, zinc, and lead that provided building blocks for the country's infrastructure and myriad consumer products. While others have explored the contribution of mining and smelting to the development of the United States, this book is about the public health and environmental costs, and the long road to taking community concerns about the health effects of smelter emissions seriously, systematically studying them, and working to protect people in smelting communities from further harm. The book focuses on three communities that were central to debates among industry, government and academic scientists, and community and environmental activists over smelters and public health: Tacoma, Washington; El Paso, Texas; and Kellogg, Idaho.

Public health and environmental disasters were discovered around smelters in all three communities after 1970, when federal and state health officials found that children who lived near these smelters were highly exposed to heavy metals from their emissions. These communities became unwittingly linked as battles over the scientific, political, and regulatory implications of these discoveries raged throughout the 1970s and early 1980s. Though the three communities are different in geography, history, and culture, there are remarkable similarities in how the disasters unfolded and particularly in the responses of the companies responsible. The conflicts and controversies that arose in smelting communities over childhood lead and arsenic exposure laid bare social and political tensions and power dynamics in each community and, more broadly, in American life.

Tacoma forms the centerpiece of the book as its public health and environmental impacts are considered across the twentieth century, providing a window into what was actually a century-long conflict between communities and industry. El Paso and Kellogg are considered mainly after 1970, with in-depth examinations of the discovery and response to childhood lead poisoning epidemics discovered near these smelters.

The smelters in Tacoma, El Paso, and Kellogg were nonferrous smelters, those that produced metals such as copper, zinc, lead and silver. One of the world's dirtiest industries, smelters have caused environmental destruction on a very large scale, harmed workers, and damaged generations of children who grew up in the shadow of their towering smokestacks. Sulfur dioxide emissions from smelters have contributed to acid rain and large-scale environmental damage. Toxic metals emitted from smelters such as lead, arsenic, and mercury do not break down; they will remain in the environment, contaminating waterways, soil, house dust, and gardens unless they are cleaned up. Though many smelters in the United States closed two decades ago or more and environmental cleanups are occurring, the potential for human exposure remains due to the vast amounts of toxic metals that have been released into the environment.

Many of the country's largest nonferrous smelters were located in the West, close to the mines from which the ore was extracted or, like Tacoma, near deepwater ports where shipments from overseas mines could be received. From its early days the smelting industry was dominated by politically connected families that organized into powerful trusts in the early part of the twentieth century. For example, William Rockefeller and Henry Rogers of Standard Oil had ownership in the Amalgamated Copper Company, which owned the Anaconda Smelter in Montana. ASARCO was founded by the Rockefellers and shortly afterward was acquired by the Guggenheim family early in the century.[2] ASARCO was the company through which the Guggenheims amassed much of their fortune, through a network of mines and smelters in the American West and around the world. The family maintained a controlling interest in the company until after World War II, mirroring the heyday of the American mining and smelting industry, which flourished between 1900 and the war. After the war, foreign competition, cheaper labor in the developing world, and other factors contributed to the industry's decline in the United States.[3]

The public health crises in western smelting communities were caused by the massive amounts of pollution smelters emitted coupled with longstanding resistance to investing in comprehensive pollution controls. Early in the twentieth century major western smelting companies made strategic decisions to rely primarily on dispersing sulfur dioxide emissions rather than capturing them, and then barely budged from this position for much of the century. To defend their stance they used science (much of it sponsored or paid for by industry), they enjoyed lax government regulation, and they threatened closure and job loss.[4] This near-century-long intransigence resulted in episodic, angry community complaints, and some smelting companies even instituted routine damage inspections and payments to individuals for dead livestock or pets or ruined gardens or laundry to try to keep the peace.[5] Damage payments, the industry found, were far cheaper than pollution controls.

The use of a tall stack to disperse considerable sulfur dioxide pollution was common in the western smelting industry, and the stacks grew taller over time in response to community ire. Even today, the rationale is a familiar one—dilution would be the solution to pollution. But the tall stack was also very effective at spreading toxic metal pollution far and wide. Around the Tacoma smelter, some one thousand square miles of soil contamination with arsenic and lead has been mapped following nearly a century of smelter emissions.[6]

For the industry, the strategy of dispersing pollution through the tall stack, coupled with assurances that health was not being harmed, worked reasonably well until the late 1960s and early 1970s, when intermittent community complaints were reinvigorated and bolstered by growing lay concerns about environmental quality and environmental health. Government at all levels gained new powers to write and enforce environmental regulations, and the federal government increased funding for independent scientific research on environmental health. The confluence of these factors focused attention on the enormous environmental and public health impact of smelter pollution in the early 1970s.

The first public health disaster discovered by government investigators was around the El Paso smelter in the early 1970s, after a lawsuit brought by the city and state over sulfur dioxide pollution led to the discovery that hundreds of tons of toxic metals were being discharged from that smelter's smokestack each year. Testing uncovered a lead poisoning epidemic among children across south and west El Paso.[7] A number of children were hospitalized, and properties inhabited by a community of low-income Mexican American smelter workers and their families adjacent to the smelter property were razed and the people were relocated.

Two years later researchers uncovered an even worse lead poisoning problem in children living near the Bunker Hill smelter. That smelter had been operating with a malfunctioning pollution control system and emitting enormous quantities of lead into the valley's air.[8] The fine particles of lead dusted homes, yards, and play areas. Children breathed it and ingested the lead-laden dust that contaminated their yards, was tracked into their homes, and wafted in through open windows.

When scientists tested area children in 1974, almost all of those living closest to the smelter had blood lead levels above 40 micrograms per deciliter (μg/dL), considered "undue absorption" at the time, and today considered a serious medical problem. Almost one-quarter had levels over 80 μg/dL—then and today considered a medical emergency even in an adult. Under worker safety guidelines, a worker with a blood lead level this high would be immediately removed from worksite exposure. One Kellogg child had a blood lead level more than twice this high—164 μg/dL.[9] A specialist in childhood lead poisoning was surprised that this child did not die.

The El Paso crisis spurred State Health Department investigators to systematically study heavy metal exposure near the Tacoma smelter. Early results revealed that some children living closest to the smelter had arsenic concentrations in their urine that were comparable to those of workers in the plant.[10] In El Paso and at Bunker Hill, the primary concern was lead, a potent neurotoxin that affects all of the body's organ systems and can cause behavioral, neurological, and cognitive problems. Although both children and adults may be harmed by the metal, children's rapidly developing brains and bodies are more susceptible to lead's damage, and the injury that lead causes to children's central nervous system is thought to be irreversible.[11]

In Tacoma the primary concern was arsenic, a notorious pollutant with a reputation for being highly toxic and lethal at high doses. In the early 1970s arsenic was known to cause skin cancer, and researchers were honing in on its role in causing respiratory cancer by studying highly exposed copper smelter workers. Arsenic exposure in Ruston and Tacoma represented a carcinogenic threat.

The sources for this book are many and varied. They include federal, state, and local government records, other archival materials, interviews, scientific journal articles in medicine and public health, newspaper articles, and court records from citizen lawsuits against smelting companies. Court records are one of the few sources of internal industry documents available to the public. Examining and analyzing them is one of the few ways researchers and the public have of determining what industry knew about the harms of its pollution and what it did or did not do to mitigate impacts on surrounding communities. Internal industry documents are usually only obtainable by lawyers in the legal discovery process. Those that wind up filed in court records represent potentially rich sources for this type of inquiry, providing access to internal discussions, industry-sponsored science, and corporate decision making that would otherwise be unavailable and unknowable to the public.

Smelters were the subjects of some of the earliest air pollution disputes in the United States. But as central as smelters are to the history of air pollution and environmental toxic metal exposure in the United States, particularly in the West, the public health story of smelting has not been told in any significant detail. Smelters are often briefly mentioned in environmental and public health textbooks as examples of significant point sources of pollution and lead exposure in children, but in-depth consideration of their public health impact is generally absent.

Most importantly, the stories and experiences of the people who lived and worked in smelting communities, raised their children in the shadow of the smelters' smokestacks, and withstood their inexorable pollution both before and after federal environmental regulation began have been largely overlooked. The significant role, at least in Tacoma, of local people in demanding that industry abate its pollution and prodding public health officials to take action has gone largely unrecognized. The contentious battles over the science of environmental heavy metal exposure and the regulatory significance of these battles have not been considered in the context of smelting communities. The fact that control of smelter emissions to the extent necessary to protect public health took decades, and in some cases never occurred, is little recognized. And the lingering public health concerns in smelting communities from legacy pollution and the long timelines for cleanup are not well understood. The ongoing threats to public and environmental health from smelting in many countries around the world have not been viewed in historical context. There is much to learn from these stories about protecting the health of people in smelting communities today.

In exploring the public health history of smelting in the United States by examining these three local cases in rich detail, the book traces an evolving argument among scientists, communities, government, and industry over industrial pollution and whether, when, and how it harms health.

Although people living near smelters were concerned about health effects from the industry's earliest days, early in the twentieth century organized community resistance to smelter pollution largely derived from economic concerns, as local people worried about pollution damage to their crops and livestock. These disputes were sometimes resolved through damage payments, "smoke easements" placed around smelters, and lawsuits. Some high-profile cases were arbitrated by court-appointed smelter commissions, which in general favored the interests of smelting companies.[12]

Science was central to deciding questions of harm, and scientific questions about the environmental and public health impacts of smelters were asked as far back as the first decades of the twentieth century. The answers had significant implications for the industry, and smelting companies had an enormous interest in weighing in on them. For the first half of the twentieth century, industry sponsored a good deal of the science, largely focused on sulfur dioxide pollution, which allowed industry to control the questions asked and the findings generated and their publication, and it allowed significant control over the public dialogue about the harms of smelter pollution.[13] This control began to erode in the postwar period as the American public became more attuned to and concerned about environmental issues and as public research funding for environmental health increased.

Public health concerns over smelter emissions began to take center stage in postwar debates over the industry's impact driven by growing public concern about environmental quality and its relationship to health. By the early to mid-1970s there was growing evidence that people living in smelting communities were being exposed to and harmed by heavy metals emitted by smelters. Smelting companies—in particular, the two considered in this book, ASARCO and Bunker Hill—used science to introduce doubt about the harms of those exposures. Working with an industry trade group, they supported studies that would challenge the findings of government research that children had been harmed by lead exposure and lead poisoning near smelters.[14]

Introducing and fostering the idea that government studies could be wrong was sufficient to spur debate and buy time as regulators considered rules on lead and arsenic in air that would profoundly affect the smelting industry. Scientist and current OSHA head David Michaels has examined this strategy of "manufacturing doubt" over questions of health effects of various environmental exposures and says the strategy has helped various industries fulfill their objectives by prolonging debates over levels of toxicants that cause harm.[15] In smelting communities, an effective response to childhood heavy metal exposure was complicated by the fact that the doubt narrative was embraced by some community residents, often those with a significant economic stake in the outcome who were disinclined to believe that the harms were real.

Even today, there are sharp divisions among residents of these communities between those who supported the companies and largely discounted the health threats and those who put stock in the findings and warnings of government researchers. There were complicated social dynamics among smelter workers concerned about their jobs; workers concerned about both jobs and health; those who generally supported industry; and those who challenged industry infringement on health, environmental quality, and enjoyment of property. The stories presented here illuminate some of the complexities of addressing community environmental health problems and demonstrate what can happen when a community's status quo is threatened by claims of health and environmental damage.

In all three communities, the companies exercised significant power, particularly Bunker Hill. Beyond sponsoring scientific studies, both Bunker Hill and ASARCO collaborated with industry trade groups (the Lead Industries Association, LIA, and the International Lead Zinc Research Organization, ILZRO), which disseminated the findings of smelter studies in the hopes of influencing environmental regulation of heavy metals, and they launched sophisticated public relations efforts to shape regulator and community opinion about harm.[16]

Public relations messages played on core American values such as independence and limited government interference in community life. Some smelting industry leaders and their supporters sought to cast government's concern for health and environment as dangerous overreaching, threatening American prosperity and innovation. Regulating industrial air pollution was cast as needless overregulation that would cost jobs.[17] In Tacoma, El Paso, and northern Idaho, ASARCO and Bunker Hill asserted that smelter pollution was not as hazardous as government scientists thought, and that children were not seriously harmed by exposure to toxic metals. These arguments were made in local communities, in statehouses, and at the federal level as the industry sought to delay and influence standards on lead and arsenic.[18]

The stories of smelting communities, though decades old, share many similarities with ongoing struggles over public health involving other industries such as the tobacco industry, where efforts to protect public health may be stymied by similar tactics.[19]

Although the three local stories are fascinating case studies in the science and politics of public health in and of themselves, all three had broader implications for federal regulation of lead, sulfur dioxide, and hazardous air pollutants like arsenic. The regulation of these air pollutants had implications far beyond the smelting industry, which was in part why the battles over the science were so hard-fought.

The cases expose tendencies in the United States to consistently prioritize economic gains over worker and community health and to wait until a disaster or catastrophe strikes or the evidence of harm is unassailable—to fail to act with precaution. These tendencies find expression in our legal system, regulatory policy, science, medicine, and public health practice, where we largely consider industrial pollution and toxic exposures safe until proven otherwise. Where health, the environment, and economics are in competition, as they often are, we engage in endless debate over possible harms while typically allowing the economic activity to proceed. Environmental health threats compete with the prevailing paradigm of health in the United States: that health is largely within the realm of control of the individual, who bears primary responsibility for his or her own health. When health is threatened by environmental exposures or industrial pollution, factors over which the individual has no control, this may be difficult for people to understand and accept as true.

These three community health disasters also provide a window into how ideas about the environment and public health have evolved and changed over time, and how researchers, government, industry, and affected people have struggled over contentions of harm. The stories also demonstrate how costly this ongoing argument has been to workers, communities, and the environment. Such struggles get to the heart of the place of health in American life.

Despite a strongly professed valuing of health in American life, the smelter story exposes how malleable this concern is—particularly when health conflicts with jobs, economic growth, and industry power. It has implications for current efforts to reduce greenhouse gas emissions—an effort that involves regulation of some of the world's most powerful corporations. This story is important to the future of the planet as we struggle to achieve a more sustainable future.

Unfortunately, the disasters discovered around U.S. smelters in the 1970s have not led to a concerted effort to stop smelter pollution from impairing children's health, development, and life chances. In fact, today lead exposure above the Centers for Disease Control and Prevention's current level of concern (5 μg/dL) still occurs near nonferrous smelters, both operating and nonoperating, in both developed and developing countries. La Oroya, Peru; several provinces in China; and Mount Isa, Australia, are just some of the places where health and environmental concerns continue to be raised both by community members and outsiders. Governments, both in the United States and abroad, have rarely forced smelting companies to strictly control their pollution for the protection of the environment or the public's health.

Continued demand for metals supports a flourishing nonferrous smelting industry overseas, and the story of environmental and public health harm from smelting is being repeated, mostly in the developing world. Struggles eerily reminiscent of those that played out here over the course of the twentieth century are now routinely reported in the international press. As advocates in far-flung places attempt to marshal scientific evidence to prove claims of harm from smelting, it is important to understand the menace that the smelting industry was to the public's health and the environment, particularly in the western United States, over the course of the twentieth century, and to understand how activists, communities, regulators, scientists, and the industry responded to the problem. It is a history, unfortunately, largely of official inaction, failure to heed warnings, refusal take community concerns seriously, denial, and sometimes outright deception. It is a history that communities and environmental regulators will be dealing with for decades to come.

1

The Tacoma Smelter

> "I don't think there was any place else in the world that emitted arsenic like this . . . and dumped it on an urban area."
>
> —Gregory L. Glass, environmental scientist who studied the Tacoma smelter for over twenty-five years[1]

Few people who live outside of Washington State have ever heard of Ruston, Washington. The town is tiny, about one square mile, and is surrounded by Tacoma. Ruston is dominated by the ninety-plus-acre former smelter site, which occupies a prime Puget Sound waterfront location. The beauty of the area—with the snow-capped Olympic Mountains to the west, Puget Sound visible in three directions, and the rolling green hills of Vashon Island to the north—contrasts with the scarred smelter site where there is little vegetation: a burial site for tons of arsenic-contaminated soil and other toxic waste.

Little known outside of the Puget Sound region, the Tacoma smelter, which operated here from the late 1800s until the mid-1980s, was one of the largest sources for anthropogenic arsenic in the world.[2] Originally built as a lead smelter in 1890, its acquisition by ASARCO in 1905 was strategic; the purchase increased the Guggenheim family's control over the western nonferrous metals industry.[3] The smelter's emissions, in addition to containing tremendous quantities of sulfur dioxide, also contained heavy metals such as lead, mercury, and cadmium, and vast amounts of arsenic trioxide, one of the most toxic forms of arsenic.

The jurisdictional distinction between Tacoma and Ruston was actually created by ASARCO in 1906, after it acquired the smelter and incorporated a town around it.[4] The arrangement made certain that the company's interests would be the controlling ones in Ruston.

Today, as in the past, many of Ruston's eight hundred residents know each other, and old wood-framed houses still line the town's streets. Once, the houses were occupied mainly by smelter workers and their families—Scandinavian, Italian, and Austrian immigrants who endured hot, dangerous, and polluted conditions to keep the smelter's furnaces burning.[5] The work was dirty and difficult,

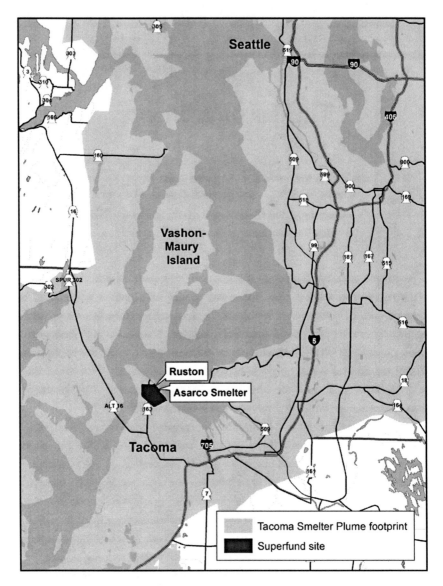

FIGURE 1.1 Map of Puget Sound Region of Washington State including the former Tacoma smelter site, Ruston, Vashon Island, and Seattle. The map shows the partial footprint of the Tacoma smelter's heavy metal contamination with arsenic and lead in the Puget Sound Region and the location of EPA's Superfund site, which includes the entire town of Ruston.

Used by permission of Washington State Department of Ecology.

and arsenic in the smelter's fumes caused skin rashes, ate holes through nasal septa, and caused workers to die of lung cancer at an increased rate.

Many, though not all, of the smelter's last workers have died, and new residents without a connection to the town's formerly dominant industry have moved in. Remaining long-time residents tend to be stubbornly proud of their industrial history. ASARCO's legacy still influences the town, though the smelter was demolished in the 1990s. Parts of the town are still undergoing Superfund cleanup, and a developer is trying to transform the former waterfront smelter site into a community of luxury residences. The development, which some see as the town's salvation and others see as its death knell, has revived old and acrimonious debates about the effects of arsenic exposure on the health of people living here.

For many decades ASARCO was the world's largest producer of nonferrous metals, operating mines and smelters in the United States, Mexico, Chile, Australia, Peru, and Bolivia, among other places. By 1939 the company owned and operated eighteen nonferrous smelters in the United States, many in the West.[6] A midcentury corporate history boasts, "ASARCO was impressive in size at birth and has grown mightily in the half-century of its existence until the entire globe is its domain."[7] The company's "tidewater" smelters, of which Tacoma was one, were located in coastal areas where they could receive ore and concentrates shipped from foreign mines in an arrangement that has been called "exploitative and classically colonial."[8]

The company is well known today in regulatory circles and former mining and smelting communities for its notorious environmental record, with responsibility for an estimated ninety-four contaminated sites in as many as twenty-one U.S. states.[9] Some of the largest contaminated sites include current and former smelting and refining sites in El Paso, Texas; Tacoma, Washington; Leadville, Colorado; Omaha, Nebraska; Silver Valley, Idaho; and East Helena, Montana.

The Tacoma smelter and its pollution problems are best understood within the larger context of the early-twentieth-century western smelting industry, and the controversies and conflicts that arose over air pollution from smelters. In the first half of the twentieth century, in the absence of much government oversight or regulation, industry needed to control smelter pollution to some extent because of citizen complaints and lawsuits. In the face of citizen legal challenges during the first decades of the twentieth century and some court orders to abate pollution, western smelting companies took some small steps to control pollution, but they did not want to practice pollution control beyond what was economically advantageous. Nominal pollution control was coupled with a strategy, which ASARCO in particular engaged in to influence the public dialogue about harm from its operations and how best to address it. Science was central to this effort, and ASARCO cultivated scientific expertise both in-house and in prestigious universities.

Industry-sponsored science helped to provide the necessary justification to allow western smelting companies to make strategic decisions to not control pollution to the extent possible with existing technology, instead opting to disperse it through a tall stack—a less expensive option.[10] Early-twentieth-century decisions to disperse rather than control pollution haunt smelting communities to this day because the strategy resulted in widespread soil contamination with heavy metals.

Dispersion of smelter pollution was not opposed by the U.S. Bureau of Mines; in fact, western smelter operators gained acceptance from the agency that, after addressing some of their most pressing pollution problems, they should not have to practice more pollution control than would be profitable. Improvements in pollution controls in the early twentieth century sometimes stemmed from directives from the courts, but sometimes came about because there was economic gain in recovering heavy metals like lead and arsenic from smoke streams. Pollution controls at many western smelters, however, fell far short of what was possible with existing technology.

Another aspect of shaping the dialogue about harm from smelting was controlling information regarding the makeup and magnitude of smelter emissions. For many decades ASARCO was extremely reluctant to divulge publicly detailed information about its emissions in Tacoma, and it promoted the idea that technological upgrades made in the early part of the century forced emissions downward to safe levels. Within the company, there was a sense of wishful thinking that the smelter's arsenic pollution was so dispersed that it would not be found in the environment, and it would not affect human or environmental health.[11] The notion that emissions were well controlled and that there was no threat to human health became a recurring theme in response to community concerns.

With some exceptions, industry and government appeared to share the goal of production and profits in the first half of the twentieth century and gave little thought to the long-term environmental or health problems associated with smelting. Claims that smelting damaged human health were mostly dismissed. In Tacoma the local government largely took a hands-off approach for many decades and deferred to ASARCO to handle the inevitable citizen complaints. It would take many decades for citizens and government to effectively challenge the narrative that smelting was not seriously damaging health and the environment.

As a copper smelter, Tacoma was part of a class of smelters called nonferrous smelters because the metals of interest are found as compounds of sulfur rather than iron. The nonferrous category includes lead, copper, and zinc smelters. The ore from which these metals are extracted contains large amounts of sulfur and impurities such as arsenic, lead, and other heavy metals. Most copper ore contains only a small amount of copper—approximately 1 percent or less.[12] Vast amounts of ore must be smelted to extract copper, producing a tremendous amount of pollution and waste in the process.

For much of the twentieth century, U.S. smelters relied on a pyrometallurgical process to produce metals—heating ore to separate the metals of interest from the other constituents. The high temperatures required for this type of smelting cause the production of tremendous amounts of sulfur dioxide—in copper smelting, up to approximately two tons of the gas for every ton of copper produced. Sulfur dioxide is associated with environmental harm such as acid rain and crop damage, and with human health effects such as impaired breathing, asthma exacerbations, and mortality. The mining and smelting process also liberates the toxic metal impurities in the ore that would otherwise remain sequestered underground. During smelting, metals like arsenic, lead, cadmium, and mercury can volatilize and emit through tall stacks. Eventually volatilized metals will cool and re-form as small particles that fall on land and water.[13]

Smelting also produces vast amounts of waste called slag. In copper smelting, for each ton of copper produced, up to three tons of slag is created, which can contain significant concentrations of toxic metals.[14] For much of the century, Tacoma's slag was dumped directly into Puget Sound and was used to build a breakwater into Commencement Bay. It has also been dispersed throughout the region, used in landscaping and road building and as ballast. Toxic metals can leach out of the slag and contaminate soil and groundwater. Fifteen million tons of slag remain at the smelter site and compose the breakwater for the Tacoma yacht club.[15]

The smelting process used at Tacoma involved four key stages. Ore was typically crushed and concentrated at the mines to prepare for the smelting process. The first step at the smelter was roasting, which drove off some of the sulfur to prepare the ore for smelting. Smelting occurred in a reverberatory furnace, in which the ore was heated to 2,552–2,732 degrees Fahrenheit to produce a molten mixture of copper and iron sulfides called copper matte. During smelting, slag forms on top of the matte and is skimmed off.[16]

The copper matte was then sent to the converters—another furnace in which further impurities were removed and the matte was "converted" into blister copper. The blister copper was sent to anode furnaces where the remaining oxygen was removed and the copper was cast into anodes. The final step, refining, drove out any remaining impurities and allowed for the recovery of precious metals.[17]

Smelting has resulted in toxic metal pollution for as long as ore has been reduced to metal. Evidence of large-scale atmospheric pollution from copper smelting dating to Roman and medieval times has been found in ice cores in

Greenland.[18] But smelting took on unprecedented dimensions in the United States around the turn of the twentieth century. The Anaconda smelter in the Deer Lodge Valley of Montana treated more ore in a week than its nineteenth-century European predecessors could have in a year.[19] Along with large-scale production came vast quantities of pollution. In the early 1900s the Anaconda Smelter discharged as much as two thousand tons of sulfur dioxide per day, and daily emissions of arsenic between 1911 and 1918 were as much as seventy-five tons.[20]

As one of the major industrial forces that shaped the development of western states, mining and smelting drove the economies and shaped the political landscape of the states in which they were located. The considerable political and economic power of the mining and smelting industry in western states like Montana, Arizona, Idaho, and Nevada persisted for much of the twentieth century. As historian Michael Malone recounts, in Montana, the Anaconda Company owned a newspaper chain and ran twenty-four-hour "watering holes" for legislators. Often with corporate headquarters on the East Coast, usually in New York City, mining and smelting companies enjoyed constituencies and political support on both coasts. ASARCO's structure provides a good example: for many years its corporate headquarters was in New York City with research divisions in New Jersey and Utah along with mining and smelting sites spread across the country. Even in the early part of the century, according to historian Donald MacMillan, there were close ties between smelting companies and the federal government, with the industry benefitting from the so-called revolving door between the mining and smelting industry and the U.S. Bureau of Mines.[21]

Toxic metal and sulfur dioxide emissions from western smelter smokestacks had dramatic effects on surrounding environments. In the early part of the century, most western smelters were located in rural or agricultural areas. The effects of their pollution that generated the most outcry were damage to crops and livestock. Soil, grazing grasses, and water sources could become so contaminated with toxic metals from smelter fallout that livestock would become ill or die and plant growth could be impaired. From 1902 to 1903, near the Anaconda copper smelter in the Deer Lodge Valley of Montana, one rancher alone was said to have lost one thousand cattle, twenty horses, and eight hundred sheep from heavy metal poisoning. The Anaconda Company paid out $330,000 in damages to local farmers in 1902.[22]

The sulfur dioxide that spewed from smokestacks could also kill or damage vegetation within a large radius of a smelter. By 1907 damage to forests was documented up to fourteen miles north, seven to eight miles to the south, and thirteen to fifteen miles to the west of Anaconda's stack.[23] Farmers and ranchers in the path of smelter pollution often saw their livelihoods damaged or destroyed by these pollutants.

Pollution damage from smelting led to some of the earliest struggles over air pollution in the United States. Early on, smelting companies often admitted to damaging crops and injuring and killing livestock and paid damages to farmers and ranchers. But what amicability had existed quickly dissolved as disputes between smelting companies and farmers were not easily resolved, damage continued, and farmers and ranchers launched nuisance lawsuits.

Both ASARCO and the Anaconda Company engaged in rancorous conflicts with farmers that lasted for decades.[24] They used similar tactics to fight such lawsuits: conducting their own research to support their positions on the effects of pollution and pollution control strategies, denying the harmful effects of smelter fumes, and impugning farmers who complained or sued, accusing them of incompetence and greed. Farmers who sued were called "smoke farmers," implying that their business was exploiting smoke damage claims rather than actually growing crops.[25]

It was in the context of these early-twentieth-century disputes that ASARCO and Anaconda began promoting taller smokestacks as the solution to smelter pollution problems. By raising the height of smelter smokestacks, they hoped to disperse stack gases and dilute their harmful effects. This thinking is described in a history of the early-twentieth-century struggle to control pollution from the Anaconda smelter: "Theoretically, as a result of discharging the smoke from the new stack . . . natural diffusion would act upon the smoke stream in a way that, by the time its poisons reached the ground, they would be so diluted as to be practically harmless."[26] Anaconda raised its stack to 300 feet in 1903.[27] Tacoma's was raised to 307 feet in 1905. The construction of taller smoke stacks was promoted with great fanfare and promises that past pollution problems would be solved.[28] The *Tacoma Daily Ledger* called the smelter's 307-foot stack one of the "most wonderful engineering feats known to the world" and reported that the higher stack would carry "away the sulphurous fumes generated in the reduction of the ores and it is believed that these obnoxious fumes will be effectively dissipated in the air."[29]

But by 1907 chemistry professors William D. Harkins and Robert E. Swain, hired by farmers to conduct an investigation around the Anaconda smelter in the Deer Lodge Valley, had determined that the taller smokestack did not solve Anaconda's pollution problems; in fact taller stacks simply made a relatively localized pollution problem more widespread, depositing toxic metals over a larger area, between one and four miles from the stack. Similarly, although sulfur dioxide ground-level concentrations near the stack generally decreased, forest damage was found farther away, where the plume eventually struck the ground.[30] Despite this early research, which accurately foretold the extensive environmental damage done by taller smokestacks, many western smelting companies would rely on them for "pollution control" through the 1970s, and some well into the 1980s.

ASARCO's Department of Agricultural Research was apparently the first in an American corporation to focus solely on air pollution research.[31] Its formation in 1914 attests to the centrality of science in struggles over pollution and associated damage claims. The company realized early in the century that the ability to shape or control research would be important to its interests.[32] The department's research played a significant role in defending ASARCO and the western smelting industry in general from farmers' allegations, particularly regarding pollution damage to crops, by promoting the tall stack and dispersion as an effective means of limiting pollution damage from smelters. In addition to conducting research in-house, ASARCO also provided research support to university researchers, notably to Robert E. Swain, the Stanford chemistry professor who demonstrated that the tall stack caused significant damage near the Anaconda smelter.[33] Swain spent much of his career studying smelter pollution and over time came to support ASARCO's position on dispersion as an acceptable means of sulfur dioxide control in the short term. He went on to lead the Salt Lake Valley Smelter Commission in 1920, and took the smelting industry's side in other such disputes.[34]

Although ASARCO experimented with some sulfur dioxide control technologies in the early part of the century, the attempts remained small-scale endeavors.[35] In general, the company viewed sulfur dioxide control at its western smelters as economically unfeasible and preferred to concentrate its efforts on demonstrating that using tall stacks to disperse smelter emissions could be effective at preventing agricultural or forest damage. Sulfur dioxide capture technology had been in use in the smelting industry since the early part of the century, specifically at the Ducktown smelter in Tennessee where operators faced legal pressure to control emissions.[36]

ASARCO's research informed the findings of smelter commissions (e.g., Selby, Anaconda, Salt Lake Valley, Trail) that were formed by the courts in the early decades of the twentieth century to mediate conflicts between the smelting industry and farmers. The commissions generally looked favorably on ASARCO's research and other industry experts and did not compel the industry to achieve maximum control of either toxic metals or sulfur dioxide. Concerns about the impact of smelters on human health often raised by smelter neighbors were largely dismissed by smelter commissions.[37]

The U.S. Bureau of Mines also demonstrated a basic acceptance of ASARCO's position that control of sulfur dioxide at many of the western smelters was economically infeasible, writing in 1917 that "for many years to come the smelters will be obliged to waste large quantities of sulfur dioxide into the atmosphere. Thus, investigations are in progress to determine how, under different climactic and topographic conditions, these volumes of sulfur dioxide, whether large or small, can be discharged into the atmosphere without causing injury to vegetation in the surrounding country."[38]

ASARCO led many investigations into sulfur dioxide crop damage on demonstration farms near the Murray Smelter in the Salt Lake Valley beginning in 1914.[39] In general, the aim of this research was to identify how much sulfur dioxide could be released from smelter smokestacks without damaging or destroying crops. Related to this, ASARCO also sought to disprove claims by German scientists that sulfur dioxide could cause "invisible injury" to plants that would lower crop yields even if the gas did not visibly burn, sear, or kill plants. American farmers in conflict with smelting industry embraced the invisible injury theory, which had implications far beyond smelting communities. In a lecture at Columbia University in the early 1920s, Robert Swain called the theory "menacing" since, "if it were sound, any industrial operation which contributed sulfur dioxide to the atmosphere was not merely a potential, but an actual agent of injury."[40] ASARCO funded research that helped to contest this theory for decades.[41] It is now well accepted that low-level chronic exposure to sulfur dioxide, which may not produce visible injury, may result in decreased growth and yield in susceptible plants.[42]

In 1917 the company's researchers put forward the idea that sulfur dioxide would only cause damage to plants at certain concentrations and under certain climactic conditions.[43] The solution to sulfur dioxide damage, therefore, was to curtail smelter operations when these conditions prevailed. At all other times, sulfur dioxide emissions should not cause injury. This approach became known as the "sea-captain" theory of smoke control because smelting operations became dependent on weather conditions and, along with tall stacks, became the standard for sulfur dioxide control at many western smelters.[44]

But as Harkins and Swain had predicted early in the century, rather than solving the problem of smelter pollution, the use of tall stacks and dispersion made pollution worse by spreading it over a larger area. Sulfur dioxide pollution from smelters wasn't just a local problem. Sulfur dioxide is a building block of acid rain, and western smelter emissions contributed to forest damage, acidification of lakes, and property damage regionally as well as hundreds of miles away. It was not until the 1970s that this cause and effect was widely recognized.[45]

The other major problem for large western smelters was how to control the vast quantities of toxic metals that spewed from their smokestacks. Early in the twentieth century, technology became available to control toxic metal emissions. Lead smelters used baghouses, large structures equipped with woolen or cotton bags through which the smoke stream passed. The bags acted as filters to capture particles of lead, arsenic, and other metals. The use of the baghouse was profitable in lead smelters because large amounts of lead could be recovered and resmelted.[46]

Cottrell electrostatic precipitators, devices that use electrical charges to remove particulates, were used in both copper and lead smelters. Captured metals such as arsenic and lead could be also be resmelted, which made the technology attractive to an industry that insisted that pollution control be profitable. The precipitator was adapted for use in the smelting industry and first used in 1907 because of the ongoing conflict between the Selby Lead Smelter on the San Francisco Bay and local farmers. The electrostatic precipitator quickly became important for recovering metals from waste gases in the smelting industry, particularly in copper smelters. Its earliest installations were at copper smelters facing court-ordered shutdown if their pollution could not be controlled—first at Selby and subsequently at ASARCO's Garfield copper smelter in the Salt Lake Valley.[47] At Tacoma, which was not facing serious legal challenges, a precipitator was not installed until 1918.

In general, baghouses were more effective at filtering metals than precipitators were, but baghouses were also more expensive to use. Their use was profitable in lead smelters because of the large amounts of lead that could be recovered and resmelted. At copper smelters, high temperatures and a higher volume of gases made their use more technically difficult and more expensive; therefore, baghouses were rarely used in copper smelting.[48]

Although both of these technologies cut down on toxic metal emissions from smelters, they did not eliminate the emissions. The industry, however, had little incentive to attempt to drive down emissions further; rather, the main aim was to capture the amount of lead, arsenic, or other toxic metals that could be profitably sold. The U.S. Bureau of Mines generally supported industry's position on toxic metal recovery, stating in 1915: "The present demand for arsenious oxide is not great enough to warrant copper smelting plants to seek a full recovery of the arsenic in their ores."[49]

There was also another problem with toxic metal control in copper smelters that relied on tall stacks and dispersion to address sulfur dioxide pollution. High temperature gas streams were needed to effectively disperse sulfur dioxide. But high temperature gas streams also caused toxic metals to volatilize, making electrostatic precipitators less efficient. At higher temperatures the devices could lose much of their efficiency and allow toxic metals to pass through the smokestack in a volatilized state.[50] Of course, once in the atmosphere, volatilized metals would reform as particles, eventually raining back down on land or water.

The trade-off between sulfur dioxide dispersion and metals control was recognized by U.S. Bureau of Mines researchers, and ASARCO scientists, in the 1910s, but it wasn't considered a significant problem.[51] Maximizing the dispersion of sulfur dioxide was thought to be more important because of the obvious injury and community outcry caused by the gas. The tall stack helped to achieve this, but it also dispersed toxic metals widely in the environment.

Until the establishment of the EPA in 1970, industry effectively won the argument on dispersing pollution through the tall stack. The tall stack combined with the practices of heating stack gases to achieve greater dispersion of sulfur dioxide and curtailing operations during certain meteorological conditions were standard at many western smelters for much of the century. By the time the federal government began to systematically examine the issue of smelter pollution in the late 1960s, western smelter operators were still relying heavily on dispersing sulfur dioxide for "pollution control." Western lead, zinc, and copper smelters were capturing only a fraction of the sulfur dioxide that their East Coast counterparts were recovering.[52] The benefit to the industry was cost savings from not having to implement effective pollution control strategies. The damage to human health and the environment from acid rain and persistent toxic metal pollution from this practice is probably incalculable.

In El Paso, ASARCO's smelter had a towering 830-foot stack. In Tacoma, as late as 1969 the company was advocating construction of a "superstack" that would rise over one thousand feet into the air to disperse sulfur dioxide.[53] The arsenic and lead that passed through the Tacoma smelter's electrostatic precipitators and was emitted through the tall stack contaminated soil in every direction from the smelter stack and was most heavily concentrated along the paths of the prevailing winds. There is evidence that Tacoma's arsenic emissions were transported as far north as Canada since arsenic concentrations at monitoring sites in southwestern British Columbia, over 150 miles north of Ruston, dropped significantly after the smelter's closure.[54]

Over its lifetime, tens of thousands of tons of arsenic are believed to have passed through the Tacoma smelter's smokestack and escaped from ground-level sources in the plant.[55] Every few years the arsenic that built up inside the massive stack would be blown out as whitish snow-like flakes and would fall down on yards, gardens, trees, and homes. The smelter's long-time physician, Sherman Pinto, described the look of the town after such an incident: "There is a snow-like appearance in the community for about three or four blocks due to arsenic discharge."[56] These arsenic showers, by Pinto's estimation, occurred about once every three years and were "followed by unfavorable publicity in the community." The contamination events, not surprisingly, would lead to the deaths of some cats, dogs, and other animals. Animals would get arsenic on their paws and skin, and they would lick themselves for relief from the highly irritating arsenic. Sometimes they would ingest enough arsenic to become poisoned. Stricken animals would die agonizing deaths or would be put down by local veterinarians.[57]

Another reason that so much arsenic passed through the Tacoma smelter's stacks is that its niche after 1912 was smelting high-arsenic copper ore.

Few other smelters could or would accept such low-grade ore, which helped to ensure Tacoma's profitability. Because it was a tidewater smelter, Tacoma was ideally situated to receive ore and concentrates shipped from abroad. High-arsenic ore that formed a large part of the smelter's feedstock came from the Lepanto mines in the Philippines and from northern Peru. Lower arsenic ore from Arizona, scrap metal, and ore from other sources was also smelted. Tacoma's ore averaged between 3 and 4 percent arsenic, although individual high-arsenic samples contained up to 11 percent.[58] The average arsenic content of ore treated at most other U.S. copper smelters was less than 1 percent.

Limited controls on pollution and what one local activist termed Tacoma's "shit ore" were responsible for very high emissions, which may have been tens of tons per day after the introduction of high-arsenic copper ore around 1913 until the first controls were installed in 1918.[59] Nevertheless, despite its size and considerable pollution, Tacoma was never the subject of high-profile court cases in the early part of the century, nor was there a commission appointed to study its pollution, as at Selby, Anaconda, and the Salt Lake Valley. Until the late 1960s, no level of government compelled ASARCO to control Tacoma's pollution, and the only real pressure it faced were the complaints of local residents.

One of the earliest recorded damage claims against the smelter was a lawsuit brought by Carl and Frederica Andersen, farmers who lived within five miles of the smelter. The Andersens alleged that the smelter's fumes wiped out their crops of fruit, vegetables, and berries, and a hive of honeybees in both 1916 and 1917. Denied compensation for their losses, they resorted to suing the company for damages.[60] A Stanford University entomologist who spent summers in the employ of ASARCO and who had conducted studies for the Selby smelter commission came to Tacoma to investigate. As an expert witness for ASARCO, he blamed the bee deaths on the poor condition of the hive and improper food.[61] The Andersens lost their case, but community concerns about smelter pollution did not go away.

The Andersens may have been right that it was arsenic that killed their bees. Arsenic is highly toxic to bees; in fact, it is notoriously poisonous to most living things, including people. Arsenic has been called the "King of Poisons" because it is a poison like no other. With the capacity to harm nearly every organ system in the body, cause cancer, and contribute to chronic disease, it has had for centuries, and it continues to have, devastating impacts on human populations. A lethal dose (70–190 milligrams) can cause a relatively violent, painful death in a matter of hours or days. Lesser amounts absorbed over time can also cause acute poisoning in which agonizing symptoms unfold over days and may also result in death.[62] The list of cancers associated with arsenic exposure includes skin, lung, bladder, colon, uterus, prostate, stomach, and liver cancers. Arsenic exposure is also associated with skin diseases (hyperpigmentation, hardening of the skin, and black warty growths called keratoses), respiratory problems, nerve damage (peripheral neuropathy), adverse reproductive effects, cognitive impairment, diabetes, and cardiovascular disease. Both naturally occurring and anthropogenic sources of arsenic have poisoned people all around the globe.[63]

Arsenic is often grouped with heavy metals such as lead, cadmium, and mercury, although technically arsenic is a metalloid, sharing properties of both metal and nonmetal. There are many different forms of arsenic, but the term "arsenic" in this book refers to inorganic arsenic, and mostly to arsenic trioxide (As_2O_3), one of the most toxic forms of arsenic. Arsenic trioxide comes primarily from two sources: it is a byproduct of copper and other nonferrous smelting, and it occurs naturally in certain geologic formations and can migrate into groundwater, contaminating drinking water. This is the tragic situation for many in Bangladesh, where a recent study estimated that one in five deaths is related to arsenic concentrations in well water of 10 µg/l or higher.[64] In parts of India, Taiwan, Argentina, and elsewhere, chronic disease and premature mortality from long-term exposure to arsenic in drinking water is a significant public health issue.

Smelting, copper smelting in particular, is responsible for the introduction of vast amounts of anthropogenic arsenic into the environment. As the smelting industry expanded in the eighteenth and nineteenth centuries in England, and around the turn of the twentieth century in the United States, arsenic trioxide began to be produced industrially in enormous quantities.[65] For example, at ASARCO Tacoma in 1953 alone, some 630 tons of arsenic was emitted through the tall stack, and between 8 and 12 thousand tons were produced for sale annually in the early 1950s.[66]

Because of its extreme toxicity, arsenic is generally regarded as an unwanted waste product at the smelter, but once large quantities are recovered, companies have a strong incentive to put it to profitable uses. Otherwise they would have to pay for stockpiling or disposal. ASARCO faced just such a dilemma in the 1950s, when it had enormous stockpiles of arsenic on hand in Tacoma. In 1953, without a market for its arsenic, ASARCO sought permission to dump it at sea. The U.S. Coast Guard, the State Pollution Control Commission, and the International Pacific Salmon Fisheries Commission all gave ASARCO the green light despite the analysis of a Wood's Hole marine microbiologist who stated that "the discharge of a material as toxic as arsenic trioxide from a moving barge at sea would be impractical without rendering a large area of the ocean toxic to marine life."[67] Fortunately ASARCO officials themselves decided that the risks of ocean dumping were too great to pursue this plan.[68]

The impetus to find profitable uses for arsenic meant that arsenic could be found in a wide array of consumer products in the nineteenth and early twentieth centuries in both Europe and the United States.[69] It was used as a green pigment in paints, wallpaper, fabrics, paper, and many other common household products, and poisoning from arsenic that resulted in illness or death was not uncommon. The poisonings, of course, affected both the workers who made arsenical products and the consumers who used them. Arsenic was also notorious for its use in homicide, as it was readily available, relatively cheap, and very effective. Mass poisonings also occurred, typically from intentionally or unintentionally adulterated food and drink. Since arsenic trioxide was sold as a white powder that had no smell, it could be confused for flour, sugar, or baking powder and could unintentionally wind up poisoning people who ate contaminated baked goods, candy, or other food and drink.[70]

Arsenic is also highly effective at killing insects and rodents. It was used as a rat poison for centuries and was used widely as an agricultural pesticide beginning in the late 1800s. Its use was problematic, however, because it could kill plants and accumulate in soils to the extent that it impaired plant growth. In 1892 arsenic was combined with lead, and the resulting compound, lead arsenate, was apparently less likely to be toxic to plants. Effective against the gypsy moth, boll weevil, and many other pests, lead arsenate was used widely on agricultural and orchard crops in the United States until World War II. Sprayers and other agricultural workers could easily be poisoned by lead arsenate, as could the consumers of sprayed fruits and vegetables. There were even some reports of death in the early twentieth century among consumers of arsenic-coated fruits and vegetables, and some physicians and health activists suspected that chronic exposure to low-level arsenic residue in food was undermining the health of the population.[71]

Other countries, such as Germany, were more judicious in their use of arsenical pesticides, which were used injudiciously in the United States until World War II, when modern agricultural chemicals began to take their place.[72] Because of its effectiveness, arsenic was "the ally of first resort whenever death must be dealt to any pest," even though farmers, scientists, and apiculturists routinely expressed concern about its toxicity.[73]

Wilhelm Hueper, who led environmental cancer research at the National Cancer Institute between 1948 and 1964, criticized the widespread use of arsenic as a pesticide in the early 1940s and called for its replacement with something less toxic that would not persist in the soil and be taken up by plants.[74] Twenty years later Rachel Carson would highlight Hueper's concerns in *Silent Spring*.[75] Both lead and arsenic are persistent soil contaminants and can still find their way into food grown in contaminated soils, even if lead arsenate is no longer in use. Recent reports of inorganic arsenic contamination in apple juice and rice produced domestically are partly due to indiscriminant use of arsenic by U.S. farmers and orchardists in the past.[76] Added to this, although the EPA phased out inorganic arsenical pesticides in the United States, the pesticides are still in use in some countries and contribute to periodic scares over arsenic-tainted food and drink.

In the early twentieth century, however, health-related concerns about arsenic exposure did not figure prominently into local calls to control smelter pollution. In fact, public clamor for smelter pollution control in Tacoma did not gain much traction until World War I, when production within the U.S. copper industry nearly doubled.[77] These production increases inevitably brought pollution increases, which provoked intense complaints and even prompted Tacoma smelter neighbors to organize the North End Improvement Club in 1916 to protest and demand damage compensation.[78]

ASARCO recognized it had a considerable pollution problem in Tacoma, and in 1916 officials were engaged in an internal discussion on how to best address it. The company's Utah Research Department thought controlling sulfur dioxide emissions should take precedence over toxic metals control. It felt certain that "the amount of arsenic that is discharged through the chimney from the copper smelting operations at Tacoma does not endanger either vegetable or animal life."[79]

But even ASARCO's research department did not advocate relying solely on the tall stack to disperse sulfur dioxide at Tacoma because of the "large amount of sulphur . . . being handled." It felt it was essential to recover some of it and even informed ASARCO's vice president that sulfur dioxide recovery in Tacoma was essential "if you wish to protect your smelting business at the City of Tacoma." Another recommendation made by the research department was for the smelting industry, including ASARCO and Anaconda, to collaborate with the Bureau of Mines and to take the issue before the Federal Department of Justice "where we could expect to obtain fair treatment, and be free from grafting smoke farmers."[80]

But company officials did not act on the internal recommendation to control sulfur dioxide. Instead, officials announced in February of 1917 that the smokestack would once again be raised—this time to 571 feet.[81] Meanwhile, no equipment was installed to recover sulfur dioxide, indicating that dispersion would be ASARCO's "pollution control" strategy at Tacoma.

Predictably, the taller stack did not solve the smelter's sulfur dioxide pollution problems. Prevailing summer winds from the north blew the smelter's emissions over the city, which brought complaints from residents whose lawns, gardens, and shrubs were burned by the toxic gas. After a fumigation in August of 1917, the *Tacoma Daily Ledger* reported that a resident sent "a bundle of the seared and browned cabbage leaves, rose bush branches, and other vegetation from his yard to the mayor." The mayor apparently sent the bundle to "the City Health Officer, Dr. Henry, who has written to the smelter officials about the smoke nuisance and received no reply."[82]

Steps to control arsenic at Tacoma first occurred with the installation of an electrostatic precipitator in 1918. Two more would be installed over the next two decades—one in 1924 and the other in 1938.[83] Community complaints and a lawsuit brought by local small-scale farmers may have played a role in the 1918 installation, but perhaps equally important were the rising prices during this time for arsenic. Although arsenic had been in use as an agricultural pesticide for many decades by this time, its efficacy against the boll weevil pushed demand from cotton growers and prices sharply upward.[84] In 1917 arsenic hit a high of twenty cents per pound, compared to five cents per pound in 1913. Prices remained high through 1924.[85] Since ASARCO faced no legal requirements to recover its arsenic, it is likely that the installation of precipitators at Tacoma in 1918 and 1924 was related to its newfound profitability.

Production increases during World War I brought unprecedented profits to the U.S. copper industry and to ASARCO. The company earned $25 million in 1916, $9 million more than the previous profitable year. Copper companies owned by the Guggenheims provided shareholders with $200 million in dividends by the end of 1918.[86] Even in this context of unprecedented profits, ASARCO did not invest substantially in pollution controls at Tacoma.

On the threshold of the 1920s the Tacoma smelter had a daily smelting capacity of two thousand tons of ore, the plant's refining capacity was expanded to five thousand tons per month, and it was being made into the second largest copper smelter in the United States.[87] Although limited pollution control was practiced, the fact that the Tacoma smelter was not located in a predominantly agricultural area may have kept it from becoming the focus of continuous and rancorous controversy similar to that surrounding other smelters. In the early decades of the twentieth century, the smelter's threat to the economic well-being of others from its pollution was typically what prompted lawsuits.

But the company still had to deal with the inevitable community complaints. A sulfur dioxide fumigation in the summer of 1927 prompted such an outcry that it led to intervention by Tacoma city officials, including the director of health, who held a meeting with smelter managers. After the meeting, Tacoma's leaders told residents that they had "no power to act," likely because the smelter was technically in Ruston, outside the city's jurisdiction. Residents were advised to complain directly to the smelter but "no relief can be expected for Tacomans from the effects of the fumes of the Tacoma Smelter except to close the smelter down, a course which it is not desired to take." Residents were reassured that there was no health threat from the fumes because the fumes were not "of such strength as to be of any serious injury to health."[88] Downplaying health effects and threatening to close the smelter in response to complaints about its pollution proved to be effective strategies; both were used repeatedly by ASARCO in Tacoma until the smelter's closure in the mid-1980s.[89]

Tacoma's claim that it had no power to act probably had little legal merit. Even though Ruston was jurisdictionally separate from Tacoma, similar cases had been won elsewhere. For example, the U.S. Supreme Court found in 1907 that the state of Georgia could force the state of Tennessee to close two smelters that were injuring its forests.[90] The Selby Smelter near Benicia, California, was enjoined from polluting a neighboring county, and the decision was upheld by the California Supreme Court in 1912.[91] Tacoma's decision not to take action against the smelter on jurisdictional grounds allowed the company to pollute with impunity for the next forty years. Over the next several decades, citizens would continue to try to convince Tacoma officials to compel the smelter to control its pollution. Although the city would take some small steps, such as establishing an air pollution control program, it would often assert that the smelter was out of its jurisdiction.

Since ASARCO was doing little to control its pollution, property damage caused by the smelter fumes was a regular occurrence. Since at least the late 1920s the company employed an inspector whose job it was to visit properties where owners alleged that smelter fumes had caused damage. The inspector would offer nominal payments to replace bushes, trees, plants, or lawns that were damaged by pollution.[92] Sometimes the company would pay to replace pets or other animals that died from arsenic poisoning.

Although the public was reassured that the use of electrostatic precipitators at Tacoma solved the smelter's arsenic pollution problem, this was not the case. As noted previously, the practice of heating the gas stream to disperse sulfur dioxide interfered with the efficacy of electrostatic precipitators. ASARCO and other smelter operators also had little incentive to recover more arsenic than could be profitably sold.[93] If too much arsenic was recovered from the smoke stream, it would have to be stored somewhere since flooding the market with arsenic would depress prices.

ASARCO kept information about how much arsenic was being emitted from the smelter's tall stack closely guarded, apparently hoping that its inevitable build-up in the soil would not be measurable. Smelter managers were even instructed by ASARCO's research department to avoid providing truthful information to government inspectors who might ask for such data. In a 1951 letter to the Tacoma smelter manager regarding a recent visit from engineers from the State of Washington Pollution Control Commission, L. V. Olson, ASARCO's director of the Department of Agricultural Research, advised the manager to be "diplomatically evasive" about stack emissions. He wrote:

> I agree that these investigators may at some future time ask for more detailed information on dust losses through the Cottrell [electrostatic precipitator], and also on the composition of the material passing through the stack. You should probably reply that the method of operating these units, for the recovery of uncommon metals, is considered a confidential trade secret, and that, of course, the Cottrell treaters are

quite efficient in removing practically all solid matter. . . . Studies have shown that arsenic discharged from the stack can seldom be found as a deposition, except occasionally nearby in large flue dust particles and it is presumed that it floats to such infinite distances and is deposited in such minute traces that it cannot be identified or found, and therefore should not be considered as an atmospheric contaminate menacing health or plant life under the conditions now existing at Tacoma. You are doing an excellent job at Tacoma in handling public relations and legislative problems. That is one place I have not worried about with these other local headaches. Your tact in handling your problems is splendid. Of course, you are aware that you should be diplomatically evasive in giving exact figures of elimination.[94]

ASARCO's position that Tacoma's precipitators essentially solved the arsenic emissions problem was uncritically echoed by scientists and government officials and in newspaper reports during the first half of the century. For example, the *Tacoma News Tribune* wrote in 1947: "There was a time when fumes caused gardeners to curse the smelly and smoky buildings but erection of what is today the second highest chimney in the world has pretty much ended this nuisance. By using a filter, arsenic in the smoke has been eliminated, and it now takes a strong north or northeast wind for residents to notice smoke that formerly covered the area with a crusty, brownish substance that killed off natural growth."[95]

University of Washington engineer Richard Tyler studied Tacoma's air pollution problems; he wrote in 1948 that "trouble has not been experienced from the chemical dusts" at the smelter since 1938, when the third and last precipitator was installed.[96] Such reassuring messages to Tacomans that the smelter's arsenic pollution was well controlled continued into the 1960s. Coupled with an official deference to the company's public statements about its pollution, an effective challenge to this idea would not come until citizens began to organize and demanded to know how much arsenic was being released into the air and whether or not it was causing harm.

Even without an organized community response, some residents alleged that arsenic emissions menaced their animals and property in the early 1950s. The ways in which the company handled these is suggestive of the power they exerted at the local level. In one instance ASARCO handled a complaint by surreptitiously testing the resident's property for arsenic residue.[97] Another complaint, which was made by an ASARCO employee on behalf of his in-laws, was solved by telling the employee that he should not blame arsenic for his in-law's sick animals "unless he were sure of his grounds."[98]

Community residents were being exposed to arsenic from stack emissions but also from so-called fugitive emissions—those that did not go through the smokestack. These included arsenic and heavy metals that escaped as ore was transported, flue dust was handled, and slag was dumped. In the mid-1970s fugitive emissions would be identified as the most significant source of arsenic exposure for residents living closest to the plant. Nearby residents were also exposed to arsenic when the precipitators periodically failed and large quantities of arsenic would be discharged from the stack, causing a "snow-like appearance in the community."[99]

Although ASARCO maintained that Tacoma's precipitators captured most of the arsenic going through the stack, there was little data available prior to the 1970s with which to evaluate this claim. However, arsenic emissions in the middle of the century were quite high based on ASARCO's own data, which put stack emissions in 1953 at 630 tons.[100]

The ability to point to authoritative scientific sources to defend themselves from complaints, the tactic of providing limited public information about its emissions, and local dependence on the company for jobs all helped ASARCO to weather the inevitable complaints about its operations in Tacoma.[101] ASARCO and researchers funded by ASARCO argued that while smelter smoke might be unpleasant and a nuisance, it did not cause health problems.[102] Local and state governments were both reluctant and ill-equipped to challenge this narrative. Such statements were also difficult for citizens to credibly question in the early part of the twentieth century since there was little independent environmental and public health research being conducted.

This began to change in the postwar period as people across the country increasingly questioned the health impacts of air pollution. As other Americans did, Tacomans turned their gaze toward their smoky skies and seriously challenged the contention that choking air pollution was just a nuisance. Episodic complaints evolved into sustained community organizing over the coming decades. Community insistence that smelter pollution represented a serious health hazard prodded reluctant local, state, and federal officials to become involved in mediating a simmering dispute that was already a half-century old.

2

City of Destiny, City of Smoke

ASARCO's copper smelter was not the only industrial plant that was fouling Tacoma's air. Tacoma, nicknamed the City of Destiny, decisively cast its lot with industry early in the twentieth century, inviting the use of its waters, land, and air for all types of industrial production with descriptions of the city's industrial potential brimming with a sense of economic triumphalism. Always in competition with Seattle, her neighbor to the north, one early-twentieth-century writer claimed the city's tideflats would "promise her the manufacturing supremacy of the Pacific Coast."[1] Tacoma did have certain advantages, including deep-water docks, an expanse of level tideflats providing access to Puget Sound, railroad lines to move freight, and, most importantly, proximity to the vast natural resources of the Pacific Northwest.

Tacoma was relatively unpopulated in 1880, but the population grew rapidly as the city became the terminus of the Northern Pacific Railroad and began to industrialize.[2] The Tacoma smelter was one of the city's largest industries, built in the late 1880s. By the early decades of the twentieth century, Tacoma's industries were churning out hundreds of millions of feet of lumber annually. Gold, silver, lead, and copper worth $12.5 million was produced by the smelter in 1912.[3] By midcentury the vision of the city's early industrial boosters was realized and the city's tideflats were crowded with industry that contributed significantly to the air pollution that plagued the city. A 1948 report mapped more than fifty-three industrial plants in Tacoma, most in the tideflats but some mixed in with residential areas. Major industries included wood products (pulp, paper, and lumber), chemical production (chlorine, soda, caustic, acid, ammonia), aluminum smelting, and ferro-alloys production.[4]

The city's railroads also contributed to air pollution by burning coal—the switch to diesel was not complete until the late 1950s.[5] The timber industry produced large amounts of wood waste, so-called hogged fuel, which was burned for fuel by local industry and played a significant role in Tacoma's smoke problem.[6] The St. Regis Pulp and Paper Company was also the target of many odor complaints. It was this plant, earlier known as Union Bag & Paper, that was often blamed for the infamous "aroma of Tacoma."[7]

In his history of Tacoma, historian Murray Morgan writes that in the early twentieth century, "few questioned the practice of discharging industrial waste into the sky for all to share. Particulate matter from the mills, like the chemical exhaust of the smelter, smelled like dollars in a community anxious for payrolls."[8] But some Tacomans questioned it, particularly those whose livelihoods were negatively impacted by air pollution. Beginning early in the century the smelter was a primary focus of sporadic citizen complaints about air pollution. But as the century wore on resistance to smelter pollution became more sustained, more organized, and more effective. Concerned community members rejected ASARCO's solution—paying individuals for damage claims—and began demanding that government regulate the industry to prevent pollution and protect public health and the environment.

The evolution from episodic community complaints to sustained organizing against smelter pollution in the 1950s and 1960s was driven by a growing belief that air pollution was not simply a nuisance but that it could have significant effects on human health. Although damage to property, gardens, lawns, homes, and cars was still of concern, in the postwar period community members increasingly linked health problems such as skin rashes, asthma and other respiratory problems, and even infant deaths to the smelter's pollution. This shift in Tacoma reflected changes occurring nationally in the postwar period as Americans became much more concerned about the environment and its impact on health.

Although citizens became more organized and vocal, the local and state government response to the problem of air pollution in general and the Tacoma smelter in particular was characterized by fits and starts and by ambivalence as government was largely reluctant to challenge industries within its borders on whom it depended for jobs and tax revenues. In the postwar period, it was becoming clear that many local and state governments were unable to effectively solve air pollution problems. By the 1960s in Tacoma and elsewhere local residents increasingly looked to the federal government to step in and either back up local efforts or take a lead role where locals could or would not.

Of course citizen challenges and organizing against industrial pollution did not go unopposed in Tacoma or elsewhere. Industry pushed back not just at the local level but also at the federal level against what it increasingly understood to be a sea change in the ways in which Americans thought about the environment and industry's right to pollute.

Concerns about the Tacoma smelter's air pollution dated to its earliest days, but by the 1930s Tacoma residents were growing increasingly dissatisfied and outspoken about the odors that emanated from other local industries as well. In 1930 the city council began an investigation of what the *Tacoma Daily Ledger* called the city's "municipal halitosis." The main focus of the investigation was the Union Bag & Paper Company.[9] But the decade of the Depression in Tacoma as well as in other U.S. cities was a challenging time to address smoke problems. In Tacoma, "survival was all."[10] Across the United States, many municipal smoke abatement efforts were scaled back due to budgetary constraints and economic considerations.[11] In the case of Union Bag & Paper, the company argued it was doing everything possible to control odors and pointedly reminded city officials: "While Tacoma has some bad smells from this plant it also has a payroll of more than half a million dollars." The company assured city officials and the public that the emissions were not damaging to health.[12]

Assertions such as these were routinely expressed by industry in Tacoma and beyond, which argued for much of the twentieth century that air pollution was simply a nuisance—it smelled bad and caused property damage—but it did not harm human health. There were few authoritative opposing voices in the first half of the twentieth century to counter these claims because industry largely controlled research on air pollution.[13]

In the 1930s Tacoma did not have a municipal program for air pollution or "smoke" control as it was called then, and it was decades behind many eastern and midwestern cities that had ordinances and programs in place by the first decade of the twentieth century. Progress controlling smoke in most cities that had such programs in place was slow—and even slower during the Depression—but historian David Stradling credits early-twentieth-century smoke reformers with laying the groundwork for more effective movements for air pollution control in the decades after World War II.[14]

By the early 1940s industries in the Northwest were recovering from the Depression by increasing production to meet demand for products for the European war effort, and Northwest cities grew as workers came west to fill jobs. Manufacturing jobs in ship and aircraft building brought job seekers from across the United States.[15] Tacoma's population topped 109,000 by 1940.[16] By this time the Tacoma smelter was the largest smelter-refinery in the United States and was producing ten thousand tons of copper each month, employing over thirteen hundred people, with an annual payroll of $2 million.[17]

The smelter already operated seven days a week, but production increased during the war. Because of restrictions on oil usage, the practice of heating stack gases to more effectively disperse sulfur dioxide was reduced, which likely resulted in higher ground-level concentrations of the gas during this period.[18]

Tacoma's victory gardeners, whose homegrown vegetables were sometimes destroyed by smelter fumes, loudly protested smelter pollution during World War II. In articulating their complaints, some residents were beginning to more pointedly and more publicly question the right of industry—particularly the smelter—to significantly impact the quality of life in the city. For example, Elizabeth Metcalf Fogg, garden club president, told the *Tacoma News Tribune* in 1942 that the "unnecessary destruction caused by the [smelter] fumes . . . [is] inexcusable. This is the general consensus of opinion of the citizens who are working with victory gardens in the area affected."[19]

In the summer of 1945 a prominent Tacoma resident and victory gardener, B. B. Busselle, told the city council that he intended to file suit against the smelter for property damage. His letter to the council, which was reprinted in the newspaper, reveals a palpable frustration about being subjected to the smelter's pollution:

> I know that thousands of victory gardeners are putting up with the same conditions and are wanting to do something about it. I for one, am going to do something about it, as I feel that the benefit the smelter may do Tacoma is being more than overcome by the damage that the (fumes) are doing to our gardens. I intend to 'instigate' a suit for damages against the Tacoma Smelter, and I am under the impression that when I do so, they will be so saturated with lawsuits that they will decide they, too, dislike the odor and poison which they are broadcasting over the community and outlying districts. This letter is long and I apologize for it being so but I am so thoroughly disgusted with the subject that were I to add the rest of the words in the dictionary I still would not have said my piece.[20]

Other air pollution problems were also of concern to Tacomans in the mid-1940s and were often discussed in local newspapers, such cattle and horses raised on the tideflats that were poisoned by fluoride emissions from the aluminum smelter. One resident railed against both the Tacoma smelter and the pulp mill (Union Bag & Paper) and warned that potential investors in Tacoma would be deterred by the fumes from both of these industries. Even U.S. military officials were complaining about the city's smoky pall when flights in and out of McChord Air Force Base, located about eleven miles south of downtown Tacoma, were hindered by reduced visibility attributable to air pollution.[21]

The growing citizen discontent with air pollution in Tacoma reflected the change occurring nationally as Americans were showing more concern about the air they breathed and the water they drank. Concerns about environmental quality in the postwar period have been attributed to a fundamental shift in the public's expectations for their standard of living, which included a clean and healthy environment. Demographic and values changes that arose after World

War II contributed to a growing emphasis on environmental quality in American life. The rise of the middle class meant that the American public became more affluent and educated, and more interested in what the environment could provide, such as clean air, water, and outdoor recreation. They also became more demanding that government play a role in providing and ensuring these environmental "amenities."[22]

The public was also growing more concerned about air pollution's health effects. Several well-publicized fatal air pollution episodes that occurred between the 1930s and 1960s contributed to this concern. Air pollution caused more than sixty deaths within three days in Meuse Valley, Belgium, in 1930. In Donora, Pennsylvania, in 1948 at least seventeen people died and thousands were made ill when heavy fog trapped air contaminants from a zinc smelter over the town. Just four years later four thousand deaths were attributed to the London fog of 1952. More recently researchers have argued that this event was responsible for up to twelve thousand deaths. These dramatic events, which were well covered in the press, provided a strong counterpoint to industry's argument that air pollution was merely a nuisance rather than a serious threat to the public's health.[23]

After the Donora deaths, increased federal funding provided the opportunity for research on air pollution's health effects that was less constrained by industry influence, and independent voices were added to the debate. Throughout the 1950s and 1960s the results of investigations into air pollution's contribution to respiratory disease, heart disease, and lung cancer were appearing in the public health and medical literature. The popular media also took up rising concerns about air pollution's possible effects on health and the environment in the postwar period.[24]

The growing focus on air pollution's health effects began to effectively challenge industry's long-held contention that air pollution was simply a nuisance. In the ensuing decades, the American public began to develop a more ecologically oriented understanding of both air and water pollution, and health concerns became even more central to arguments for stricter controls on pollution.[25]

In the 1950s and 1960s, however, despite a burgeoning concern for the environment, controlling air pollution remained a challenge. Regulatory authority for air pollution control was still in the hands of states and localities, and the federal government was reluctant to assume enforcement power over what was largely viewed as a local and state problem. Little progress was made in the 1950s and 1960s in cleaning up the air over most urban centers, which led to growing calls from the public for the federal government to play a more active role.[26]

Demographic changes in Tacoma, which included a growing middle class not dependent on the smelter for jobs, as well as growing concerns about air pollution's health effects had much to do with sustained community organizing that arose in the late 1950s. Residential development in the mid-1950s brought middle-class homeowners directly into the path of the smelter's smoke stream. When their entreaties to ASARCO to abate its pollution did not solve the problem, concerned residents began organizing and demanding that government at all levels take action to protect public health. Although residents were motivated both by quality of life and health issues, they framed smelter pollution primarily as a health issue.

Air pollution disasters, particularly at Meuse Valley and Donora, were also a catalyst for community organizing in Tacoma. The disasters had particular salience locally because one of the main air contaminants implicated in both the Meuse Valley and the Donora disasters was sulfur dioxide pollution from smelters. In public discussions of Tacoma's pollution, citizens often mentioned the deadly disasters that had occurred elsewhere to bolster their claims that Tacoma's pollution was affecting their health.

In Tacoma, community insistence on government intervention was the impetus for government's subsequent reluctant involvement, which included funding for independent research. Increased citizen and government involvement began to erode ASARCO's authority to define the scope and importance of the problem as well as its possible solutions.

Prior to the sustained community organizing against the smelter that emerged in the late 1950s, the city of Tacoma had taken some halting steps toward controlling industrial air pollution. The first tangible step occurred in 1947, when the Tacoma Engineers Club, a group of city and industrial engineers, proposed that the city adopt a smoke ordinance recommended by the American Society of Mechanical Engineers. The group cited health and quality of life benefits, fuel savings, and safer aviation as reasons for taking action.[27] But the adoption of an ordinance in Tacoma proved to be contentious because industrial pollution control was viewed as conflicting with economic growth. In the weeks preceding a planned meeting with city leadership to discuss the Engineers Club proposal, the mayor emphasized: "Nothing will be done to upset this city's economy in favor of any plan to combat the smoke nuisance."[28]

But Tacoma's air pollution problems were serious and researchers were just beginning to document them systematically. A 1948 report noted that "tourists and others driving through Tacoma get the impression of excessive smokiness or smog. The odor from the pulp plant has added to the impression that the city is primarily a manufacturing center and not an attractive residential city."[29]

The more than fifty industrial plants in the city were largely to blame for the city's poor air quality. Cars and home heating sources made lesser contributions. Reflecting a growing tendency to view air pollution as a health threat rather than a nuisance, the 1948 report linked lung problems, infant mortality, and an increased death rate to urban air pollution—conditions for which the evidence for the role of air pollution has grown much stronger in recent decades.[30]

Tacoma's industries would fight hard against legislative efforts to achieve real pollution control, and they organized their opposition through the Tacoma Chamber of Commerce's Industrial Bureau. Interestingly, the chamber was also responsible for prodding ASARCO to take the first step toward controlling sulfur dioxide emissions by installing a sulfuric acid plant at Tacoma. As the spokesperson and advocate for Tacoma's industries, the chamber was on the receiving end of citizen damage complaints caused by smelter fumes.[31] In the late 1940s the chamber's efforts paid off: ASARCO took the first step to capturing sulfur emissions after determining that there was a market in the Puget Sound region for sulfuric acid.[32]

The chamber's role in pushing ASARCO to invest in pollution control seems to contradict its role in attempting to weaken smoke control efforts by government. Its interest, however, was in maintaining a favorable business climate in Tacoma, which could only be helped by mitigating such a serious air pollution problem. Additionally, pointing to voluntary action on the part of local industry to address pollution could help to forestall regulation. The acid plant that ASARCO installed in the late 1940s removed only 18 percent of the sulfur dioxide generated and was not always fully utilized.[33] It did not prevent fumigations, property damage, and resident complaints.

In 1950, three years after it was first proposed, Tacoma passed a smoke ordinance, with strong support from the Pierce County Medical Society, which pointed to the Meuse Valley and Donora disasters as justification for its passage. The city's ordinance, however, was limited to regulating dust and black smoke—visible emissions—precluding a focus on the more complex mixture of visible and invisible air contaminants made up of gases, chemicals, and particulates, increasingly being referred to as "air pollution." The law also provided relief for established industries "which would be unduly burdened" by compliance.[34]

There are differing accounts of how the ordinance was developed, but the Tacoma Chamber of Commerce played a lead role.[35] Its passage was announced with city officials and business leaders lowering the public's expectations for quick relief from pollution and with the mayor trying to temper the zeal for reform by stating: "[We] must all remember that the city is basically supported by industry."[36] The mayor's statement encapsulated the ongoing difficulty and ambivalence of the city's leadership in cleaning up Tacoma's skies. Although elected officials might have liked the city's air to be cleaner, in the face of powerful industrial opposition, this would take many decades to achieve.

The city's efforts, however, were of little consequence to ASARCO. The smoke ordinance largely did not apply to its emissions; at the same time, the company did not have to worry about Washington State government taking a strong stance against industrial pollution. In 1950 the state had only begun to study the problem and consider preliminary legislation.[37] And a year later the state decided not to move forward with air pollution control legislation for at least a few years on the advice of an advisory committee in which Tacoma's industries were well represented by Weyerhaeuser, the Tacoma Chamber of Commerce, and the Hooker Chemical Company. The advisory committee did not want statewide regulations to discourage industry from locating in Washington and instead recommended "study-type" legislation that promoted collaboration between universities and industries to encourage the resolution of air pollution problems at the local level.[38]

Citizen calls for a stronger government role in air pollution control in the late 1940s and early 1950s resulted in a backlash on the part of polluting industries. By the late 1940s the Manufacturing Chemists Association (MCA), an industry trade group, organized an air pollution abatement committee that tracked local, state, and federal legislative developments and tried to alter, weaken, or defeat proposed legislation it found "objectionable."[39]

An ASARCO consultant told a 1952 MCA gathering that industry should use a two-pronged strategy in fighting legislative developments: "protect itself against unwarranted charges and at the same time improve its public relations program to stave off restrictive legislation on air pollution."[40] In the ensuing decades, public relations would become key to ASARCO's efforts to counter public fears in Tacoma that health was being damaged by the smelter, to downplay concerns about environmental pollution, and to fan the city's anxieties about job losses.[41]

Tacoma's smoke control program struggled to make a dent in the city's pollution problems in the early 1950s, but it was never well funded, nor could it overcome industry co-optation and the ambivalence of the city's political leaders. By 1953 the smoke control officer left to join private industry, and the *Tribune* reported that he had done "the best he could with limited resources and in the face of some powerful industrial opposition."[42] By the end of 1953 the program was all but dead.

Meanwhile, smelter emissions continued to plague the north end of the city. Even with the new sulfuric acid plant, ASARCO's inspector would report that damage payments in 1952 exceeded the amount paid in any other year—by August ASARCO had paid $11,000 to compensate residents for damage to gardens and vegetation. The company sensed the growing public intolerance for industrial pollution. Comparing a severe summer fumigation in 1952 with one in 1927, ASARCO's inspector wrote: "We live in a far different world in 1952 than in 1927. Now complaints are mountainous that were mere mole hills in those days. How I wish my strong recommendation of 1927 that we buy up all of that cut-over stump land had been acted upon!"[43]

Local government efforts faltered in the fall of 1953, but the public did not stop complaining about the pollution blanketing the city. With Los Angeles' brown haze now being called "smog," a *Tribune* editorialist suggested that Tacoma come up with a unique name for its problem—"FODOR"—a blend of fog and odor.[44]

Just as Tacoma was having difficulty overcoming its inertia on air pollution, so was the state of Washington. The U.S. Public Health Service (USPHS) inquired into the state's programs in 1954 and received this blunt response from the health department: "State activities in atmospheric pollution are so fragmentary in the state of Washington that very little worthwhile information can be supplied." The health official was unaware of any state-level legislation under consideration.[45] Confronting air pollution meant a direct confrontation with industry—something that neither the state nor the local government had the appetite for in the 1950s.

Throughout the decade of the 1950s the State Department of Health, within whose purview air pollution fell, remained skeptical of air pollution's health effects, and immobilized by its concerns about industrial opposition to state regulatory efforts. Health officials preferred that the department confine its role to conducting surveillance, providing technical assistance to local health departments, and working cooperatively with industry rather than assuming a regulatory role.[46]

Probably due to ongoing industry opposition, the city of Tacoma's focus on industrial air pollution in the mid-1950s began to include an emphasis on contributions made by individuals and households, with "automobile exhaust, backyard incinerators, and fireplace smoke every bit as much trouble as the big industrial plants on the Tideflats."[47] Although it was undeniable that individuals were contributing to Tacoma's dirty air, industrial emissions were the predominant source. With fewer than 155,000 residents in 1950, the city did not have the population density of midwestern and eastern cities, nor did it rely on coal as a primary fuel for residential heating by this time. Tacoma also did not have to worry about pollution from elsewhere settling over the city. For much of the year, prevailing winds bring southwesterly air from over the Pacific Ocean, which at that time was very clean.[48]

A 1956 report estimated that Tacoma's motorists, driving one and a half million miles per day, put less than one ton of sulfur dioxide into the air. In comparison, the Tacoma smelter, operating at full capacity, was emitting an estimated twenty-two tons of sulfur dioxide per hour in 1970—one of the earliest years for which figures are available.[49] Smelter emissions in the mid-1950s may have been even higher, but city officials publicly downplayed the problem. A new smoke control officer had been hired by this time and told the *Tribune* in 1956: "Generally speaking high concentrations of sulphur dioxide—which can lay waste to vegetation—are almost non-existent. But occasionally, when the wind is blowing just right, folks in the West End of town walk around all day with a sulphur taste in their mouths."[50]

By the mid-1950s it was inevitable that the western portion of Tacoma's North End, hereafter referred to as the West End, would become home to more residents. Tacoma was growing: its population increased by about thirty-eight thousand people between 1940 and 1960.[51] Land described by ASARCO's damage inspector as "cut-over stump land" that the company had once considered buying up was being transformed into desirable middle-class neighborhoods in the 1950s and 1960s. The new residents would not be as dependent on ASARCO for their livelihoods, which would influence their response to the smelter's pollution.

Despite the impact of the smelter on area vegetation, the West End was a place of striking natural beauty, its hillsides sloping down to Puget Sound, that offered, on a clear day, views to the west of Puget Sound, the Tacoma Narrows Bridge, and the Olympic mountain range, and to the north, the fir and cedar forested hills of Vashon Island. It was unavoidable that the development would create conflict between the new residents and ASARCO since the neighborhood was directly in the path of the smelter's smoke stream during summer months.

Well aware of the planned development, ASARCO watched the changes in the West End with alarm. Robert E. Shinkoskey, the smelter manager, served on a local planning committee and kept ASARCO's leadership in New York City updated on local decisions. Company officials clearly understood that adding more residents to the area would subject more people to the smelter's noxious fumes and would lead to ever-increasing complaints. Referring to the development of two thousand acres under discussion by January 1956, the smelter's damage inspector informed L. V. Olson, ASARCO's director of agricultural research: "This area is in the general direction of the smoke stream when fair weather prevails. Obviously, there is nothing we can do to prevent homes being built in this area. I am sending it to you to point out the growing hazards that we face."[52]

In the mid-1950s Tacoma's West End was home to about twelve thousand people. A planned two-thousand-acre development was expected to add over forty thousand more residents. In the winter of 1957 Shinkoskey warned Olson that the area was "in the heart of our smoke stream" and was "our area of greatest fumigation." He concluded: "Our smoke problems will be compounded if the expected growth of this area materializes."[53] Drainage and inadequate roads stood in the way of West End development, but these issues were addressed in the mid to late-1950s, making way for thousands of new homes.[54]

One of the West End's new residents was Judy Alsos, a native of Australia who, with her husband, bought a parcel of land in 1955 in the Skyline neighborhood, about three miles to the southwest of the smelter. New to Tacoma, the family did not know that it would be living downwind of a smelter during the summer months. Looking back on the day she first visited the property, Mrs. Alsos said that at the time she thought it was odd that there was very little vegetation in the area. A madrona tree was all she could recall.[55] A study conducted in the 1970s would add scientific validity to this casual observation. Researchers found that smelter emissions had impacted natural vegetation within four to five miles southwest of the smelter, which would have included the Alsos property.[56]

The Alsoses built their house and moved in with their two young boys. Shortly thereafter they became aware of sulfur fumes in the neighborhood, which they experienced as more of a taste than a smell. It didn't take long to determine the source since the smelter's smokestack was visible from the front yard. Judy Alsos recalled that neighborhood children experienced coughs, trouble breathing, asthma exacerbations, and skin rashes when playing outside. On particularly bad days a neighbor would drive her asthmatic child across the Narrows Bridge and out of the path of smelter pollution to prevent an attack. Judy's own son would sometimes wake up at night with difficulty breathing.[57]

The respiratory symptoms recalled by Judy Alsos are consistent with sulfur dioxide exposure, although smelter neighbors did not have the validation of the large body of scientific evidence on health effects that has been developed in the intervening decades. Breathing the gas can cause constriction of the airways, mucus secretion, cough, shallow breathing, and cardiovascular effects. Infants and children are highly susceptible to air pollution's effects in part because their lungs are still developing. Asthmatics are particularly at risk because exposure to short-term concentrations of sulfur dioxide can cause asthma attacks. Increases in mortality have also been reported following short-term peak exposures to sulfur dioxide.[58]

Although the research on sulfur dioxide's health effects was in its infancy in the 1950s, there were some studies that implicated sulfur dioxide in causing human health damage. Some significant work in this area was actually funded by ASARCO and conducted at Harvard University under the direction of Philip Drinker, who, in addition to his Harvard University appointment in the Department of Industrial Hygiene, was the first director of ASARCO's department of hygiene. Drinker was a central figure in the early years of industrial hygiene research in the United States and an influential scientist. One of Drinker's areas of expertise was sulfur dioxide research, for which he received funding from ASARCO throughout the 1950s. Scientist Mary Amdur, who worked under Drinker at Harvard University in the early 1950s, conducted ground-breaking research on the effects of sulfuric acid and particles on human and animal lungs. In sum, her work was among the earliest to show that lower-than-lethal doses of sulfuric acid (produced when sulfur dioxide combines with water in the atmosphere) combined with particles could cause lung damage in animals and people. At a scientific meeting where she was to present her findings, Devra Davis recounts that she was harassed on an elevator by "two large, tough-looking guys wearing leather jackets." A scientific paper reporting on the findings had been accepted by the *Lancet*, but Davis reports that Drinker "told her to withdraw the paper.... When she refused, her position with Drinker was eliminated."[59]

Sulfur dioxide's ability to damage vegetation was obvious to the smelter's neighbors even though experts at the time argued over the concentrations that caused damage. Maintaining a lawn or a garden could be difficult in the West End. In the summer, plants, trees, and lawns could be easily burned by sulfur dioxide fumigations. When the pollution was particularly bad, residents would call the smelter to complain. Managers would sometimes respond by curtailing production—a time-consuming process that did not provide immediate relief. These conditions combined with reports of infant deaths in the neighborhood in December of 1957, which some believed were linked to smelter pollution, motivated residents to take action.[60]

The belief that sulfur dioxide was contributing to infant deaths was not totally implausible since a 2004 study found an association between sulfur dioxide air pollution and sudden infant death syndrome, at average concentrations far below peaks recorded in Tacoma in the 1950 and 1960s.[61] Although it is difficult to know with certainty the concentrations of sulfur dioxide that North Enders were routinely exposed to given the smelter's emissions, there is every reason to believe that they could have been very high in the 1950s and 1960s depending on wind direction and weather conditions. An internal ASARCO report indicated a half-hour peak of three parts per million in the summer of 1952, and ASARCO admitted to concentrations this high in the community in 1961.[62] In the early 1960s the city began limited sampling of sulfur dioxide air concentrations. Although it did not publicly report the data, a citizens' group was told that on occasion sulfur dioxide reached the limits of the equipment's detection (five parts per million).[63] Peak concentrations of three to five parts per million measured in Tacoma's air in the 1950s and 1960s would have been forty to sixty times higher than EPA's recently promulgated short-term sulfur dioxide standard would allow.[64]

Also contributing to local concerns about smelter pollution were frequent media reports on the health hazards of air pollution. A May 1957 article that appeared in the *Saturday Evening Post* titled "That Creeping Menace Called Smog" showed photos of Donora victims being taken to the hospital on gurneys, citizens wearing "improvised gas masks" in Los Angeles, and women wearing nylon stockings that allegedly disintegrated on contact with sulfuric acid in the air of American cities. The article warned readers: "What happened to Donora could happen to almost any industrial community in the same unlucky circumstances."[65] At a meeting of Tacoma residents concerned about smelter fumes, the *Saturday Evening Post* article was brought out for reference.

For Judy Alsos 1957 would mark the first of more than ten years of advocacy and urging various levels of government to impose controls on the smelter. She first took her concerns directly to local smelter managers and ASARCO president Charles Barber, informing him that she would get the support of garden groups and city and county officials if the smelter did not abate its fumes. Unsatisfied with the outcome of these efforts, Alsos and some of her neighbors called a community meeting to discuss forming a group to protest smelter pollution. ASARCO learned of her organizing efforts through Tacoma's city manager David Rowlands.[66]

The large number of damage claims and growing resident dissatisfaction led L. V. Olson, ASARCO's director of the Department of Hygiene and Agricultural Research, to make a trip to Tacoma in May of 1957 to investigate the situation. He arranged a personal meeting with Alsos, who had previously refused the company's offer of thirty dollars to compensate for damage to her property. His extensive notes from this visit to Tacoma provide an account of resident organizing efforts, including his observations of a meeting organized by Alsos and her neighbors that drew at least seventy residents, many of whom invoked protecting their health as a reason for stopping smelter pollution. When neighbors pointedly questioned him about sulfur dioxide's health effects, Olson tried to assure them that no health threat existed by pointing to smelter workers who were routinely exposed to high concentrations of sulfur dioxide "continuously . . . with no ill effects." Despite his reassurances to smelter neighbors, even Olson thought that the company should take steps to address pollution concerns in Tacoma, arguing internally, "concentrations of gas are too high."[67]

Out of community meetings came a West End grassroots group called the Citizens' Committee on Air Pollution with Alsos as its president. The committee went from knowing little about the local smelter to understanding that it was owned by one of the most powerful mining and smelting companies in the world. It used the Tacoma Public Library to school itself on smelting technology, air pollution control, and the economics of metals markets. It studied the scientific literature on the health effects of sulfur dioxide pollution and became conversant with the research going on at the time. It also became skilled at communicating with elected officials, using the media, and giving testimony at public hearings.

With citizen complaints growing and the Citizens' Committee pressing local government for action, the city of Tacoma applied in September of 1957 for a federal Community Air Pollution Demonstration Project to measure sulfur dioxide concentrations in the city. The State Department of Health provided a strong endorsement for the project, writing that Tacoma's sulfur dioxide problem was "suspected to be of substantial proportions."[68] Up until this time the city had no ability to measure sulfur dioxide in air and limited capacity to measure air quality in general.

City Manager David Rowlands and Smoke Control Officer Joel Durnin were in the difficult position of mediating between citizens and ASARCO. Alsos remembered Durnin as sympathetic to residents' concerns about smelter pollution but unable to produce tangible results: "He knew that something was terribly wrong, and yet he was between doing for us and losing his job at the City. That's what it amounted to."[69]

In 1960, in response to the complaints he was receiving from residents, Durnin attempted to survey residents about their experiences with air pollution, aiming to assess property damage and "physical discomfort." He developed a mail survey, but before he could send it out, ASARCO learned of its existence and immediately began to take steps to stop it, working through its allies in the chamber of commerce, local industry, and elected officials. The incident provides insight into the level of control that ASARCO had over local air pollution control efforts as well a concrete example of how it exercised its influence. An internal ASARCO document explained that Mr. Shinkoskey (the smelter manager) tried to contact Durnin's boss, David Rowlands, but found him out of town. Shinkoskey then contacted Durnin directly and "told him that if this questionnaire is sent out, it would have a very serious effect on industry in the area." Durnin promised not to send out the questionnaire until after he checked with David Rowlands. Tacoma smelter managers also called three industry members of the City Advisory Board on Air Pollution regulations to get their support. Two pledged to "take a stand against Durnin's action," and one offered to call Durnin and advise him not to send out the survey without consulting the board. ASARCO also contacted the chamber of commerce secretary, E. R. Fetterolf, who "said he will see the City Manager Rowlands on his return Monday to point out what a very serious action this is." Its final step was to call "a meeting of the Air Standards Committee for Wednesday, March 2nd to consider this matter."[70]

Top ASARCO officials in New York City were made aware of the planned survey and weighed in on the importance of scuttling it. ASARCO vice president Edward Tittman wrote to Shinkoskey and urged him to "leave no stone unturned in heading off this type of an investigation."[71] Had Durnin succeeded in collecting data and documenting citizen concerns, this would have been the first time that data had been collected on the effects of ASARCO's pollution.

In the end, with the help of the chamber of commerce, ASARCO successfully prevented the survey from being sent out. Shinkoskey reported on how this was accomplished to Vice President Tittman:

When Dave Rowlands, Tacoma City Manager, returned February 29th I talked to him at some length about Durnin's proposed smoke questionnaire. I told him I couldn't see that any good would come out of such a survey, and as a matter of fact, it would probably stir up a hornet's nest. He is aware that ourselves, Ohio Ferro-Alloys, St. Regis, U.S. Oil and the two major Chemical companies have a common problem and that we are all doing a good job of smoke control. Rowlands also understands there are definite limits to what can be done short of shutting down. Rowlands has proven to be very helpful in the past, and he said he would discuss this with his Advisory Board before taking any action. E. R. Fetterolf, Secretary of the Chamber of Commerce, had lunch with Rowlands Tuesday, and Rowlands stated unequivocally "the questionnaire will not be sent out." Durnin reported to our local Committee on Air Standards today he "couldn't get any funds to send out the questionnaire." For the present it appears we are over this hurdle, although knowing Durnin, I don't think we have heard the last of him.[72]

The incident is illustrative of why industry favored local control of air pollution. Power and influence could be wielded easily at the local level where city and county governments were dependent on industry for tax revenue and jobs and where industry leaders had personal relationships with local government officials.

Around the time they were working to quash Durnin's air pollution survey, ASARCO and other Tacoma industries were also fighting a move in the state legislature to establish a state air pollution control board that would have the authority to set air quality standards. At a hearing on the proposed bill in Tacoma in March of 1960 ASARCO testified that such a bill could not solve the smelter's air pollution problems since the company had already done everything possible to control emissions and that one of the few options for improvement was closure of the smelter.[73]

By November of 1960 Durnin apparently had abandoned his efforts to address the smelter's pollution. Quoted in the *News Tribune*, he said of Tacoma's air: "There is no really serious problem existing in air pollution today which is under our jurisdiction," reverting to Tacoma's longstanding position on the smelter, that the city had no power to intervene.[74]

Although Tacoma city government had effectively been silenced for the time being, local residents were about to become more vocal in their opposition to the smelter's pollution. In the early 1960s two "arsenic showers" occurred in which fallout from the tall stack rained over the West End. ASARCO admitted to the arsenic showers and said that they were probably caused by the "sloughing of incrustations from the stack."[75] While so-called stack showers occurred periodically, these may have been the first, or the most noticeable, to occur since the West End had become populated with new residents. Judy Alsos recalled that a neighbor lost three puppies in one of the pollution events.[76] These incidents reinvigorated the Citizens' Committee's efforts to put an end to smelter pollution. This time it mostly focused its efforts at the state and federal levels after having had little success with local government. In a statement to the governor of Washington, the Citizens Committee wrote: "We would like to state that we do not feel that air pollution control is workable at a local level. In Tacoma our air pollution control is practically non-existent."[77]

Increased citizen scrutiny and public questioning of the content of the smoke stream was important for beginning to understand the nature and extent of pollution in the community. Always a critical piece of information for community residents, good data on actual emissions from nearby industrial facilities can be difficult for citizens groups to get even today. The Citizens' Committee took the lead in trying to determine the makeup of the discharge that had rained down on the neighborhood. It collected a sample of the fallout and took it to a laboratory in the nearby town of Fife. According to Judy Alsos, the lab refused to analyze the sample since the smelter was a client of theirs, so the committee took the sample to City Hall and pressed to have the sample analyzed. The city in turn asked the state to do the analysis, but the state was reluctant to get involved and preferred to take ASARCO's word for it that it would try to prevent future fallout episodes. Additionally, the State Health Department said, "In view of our critical budget situation, we are not in a position at this time to check the results of such further control steps by field sampling and measurement."[78] By October of 1961 city officials had tested smelter fallout that was found on a car. The sample contained 3.5 percent arsenic by weight.[79]

The health effects of environmental arsenic exposure were not well understood in the early 1960s when Tacoma officials found that it was being dumped on their city, probably in very large quantities. Arsenic's role in causing skin cancer had long been suspected and was generally well accepted in the medical and public health literature at this time. There was also growing speculation that arsenic might be a respiratory carcinogen. However, chronic community exposure to inhaled arsenic had not been studied; in fact, few researchers had even posited that such exposure could result in health effects. So when the Citizens' Committee stepped up its efforts to lobby government to address the smelter's pollution, there was little in the scientific literature to suggest that the Tacoma smelter's arsenic pollution was impairing community health.

In the summer of 1961 Alsos and the Citizens' Committee presented a petition with 1,150 signatures opposing the smelter's pollution to the Tacoma City Council and the Pierce County Commissioners.[80] The petition caused the city council and the county commissioners to agree to sponsor talks with the smelter and the Citizens' Committee on the pollution problems. The talks did not produce tangible outcomes, so the committee next appealed to the governor of Washington. The appeal prompted the State Department of Health, for the first time, to ask ASARCO what it was emitting and in what quantities—in effect, positioning itself as mediator between the community and the company. The health department, however, was careful to represent its intentions as benign, writing: "It is our desire to see industry, agriculture, and the people living and progressing amicably in a mutually acceptable atmospheric environment."[81] The agency's conciliatory, noncommittal language reflects the tenuous position of local and state governments, which increasingly found themselves betwixt and between daunting industry opposition to pollution control and increasingly vocal community protest.

While smelter officials publicly indicated an openness to share emissions data with government officials, and even told Washington senator Henry Jackson's office that they made the data available "upon request by authorized public officials," this was not the perception of city and state officials.[82] Dr. Bernard Bucove, the State Health Department director said in a 1961 briefing to the governor that although ASARCO had done more research on sulfur dioxide emissions than any other firm, "they have been rather secretive about their results and progress of their research."[83] Durnin said in 1967 that the company had been "reluctant" to share the data it was collecting at seven sulfur dioxide monitoring stations positioned around the city.[84] At this time neither the state nor the local government had the legal authority to compel the company to hand over its emissions data for inspection.

At Governor Rossellini's request, the Citizens' Committee first met with state and local health department representatives to see if the dispute could be resolved without the governor's involvement.[85] Alsos told the health officials that both sulfur dioxide and arsenic were harming the health of local residents. Dr. Bucove expressed concern but said that the state legislature's failure to appropriate funds for air pollution control was keeping the department from getting more involved.[86] Not satisfied, the committee pressed for a meeting with the governor.

The meeting with Governor Rossellini was scheduled for early December 1961. In a briefing for the governor, Dr. Bucove wrote that the smelter problem was a "longstanding" one and that the committee would want to discuss with him the legislature's failure to fund the state's air pollution program as well as some infant deaths in Pierce County "which may have been contributed to by the air pollution in that area."[87]

When Judy Alsos and the Citizens' Committee traveled to Olympia to present the governor with a petition signed by thirteen hundred smelter neighbors, they were accompanied by a group of young activists who sang protest songs and burned sulfur on the steps of the capitol building. Alsos called the smelter's pollution "out of hand" and said that people in the community were experiencing respiratory problems from sulfur dioxide, skin rashes, and deaths of animals from arsenic.[88] She reminded the governor that sulfur dioxide was "the culprit in most all of the known air pollution disasters," including those in Donora, London, and the Meuse Valley. The symptoms experienced by Meuse Valley sufferers were said to be similar to what she and her neighbors experience when sulfur dioxide fumigations occur—"throat irritations, hoarseness, cough, shortness of breath, a sense of restriction of the chest, nausea, and some vomiting."[89]

Rebutting ASARCO's claims that pollution controls were not economically feasible, she alleged that the company's profits doubled between 1959 and 1960, yet it was not spending enough to control pollution in Tacoma. She asked the governor to discuss the problem with ASARCO's New York management with the Citizens' Committee in attendance, and to find a solution.[90]

In response Governor Rossellini said, "I think all of us would like to see pollution of our air corrected. We are now about at the point where we can start to move." He pointed to the newly formed State Air Pollution Control Board as the appropriate forum for addressing the problem. Both Judy Alsos and ASARCO officials were invited to discuss the problem with the board. But the board had no real power to compel industry to control its pollution. Created by a 1961 amendment to the state's air pollution law, nine board members representing industry, agriculture, labor, local government, and the state were to oversee the air pollution control activities of the State Department of Health, but they had no enforcement powers.[91]

These small steps taken by the state were meant to appease public concerns about air pollution, but they did not position the state of Washington to take decisive action. At the same time, the State Department of Health was displaying more concern about the smelter problem and was supporting some of the Citizens' Committee's recommendations for pursuing a course of action, such as bringing in outside experts to evaluate the problem and establishing continuous monitoring of air quality around the smelter. But the department felt that its budget would not allow for "substantial monitoring of ambient air quality" or community health studies.[92] By the early 1960s the State Department of Health had just 2 full-time positions devoted to air pollution control, leaving it with neither the financial nor human resources to fulfill even a limited technical assistance role to localities. And until 1960 there were no staff with the medical expertise necessary to study or analyze air pollution's health effects.[93]

Just as the Citizens' Committee was trying to mobilize high-level political support for its position, so was ASARCO. Charles Barber, the company's president, wrote to Washington senator Henry M. Jackson's office in November 1961 to inform him that the company "always recognized our obligation to reduce and clean smelter gases to the extent economically practicable for the protection of our workers and the residents of the surrounding community." Further reductions in Tacoma's pollution, however, were not economically feasible. Barber noted: "We have been reassured by the recognition on the part of public authorities that the problem is one to be dealt with by a balancing of interests. All concerned have recognized the importance to Tacoma of the continued operation of the plant."[94] A copy of the letter was also forwarded to the governor's office.[95]

Around the same time that the company was assuring high-level elected officials that it was doing everything feasible to control its pollution, top officials were internally debating whether to spend the money to expand Tacoma's capacity to recover sulfur dioxide. Tacoma manager Shinkoskey argued against expansion on the grounds of cost, writing, "I don't think the present smoke situation is serious enough to warrant spending several million dollars to expand our acid plant. With the poor business outlook for the Tacoma Plant, I cannot recommend any large expenditures now."[96] L. V. Olson, however, felt that ignoring the problem was not an option; in October of 1961 he wrote, "Considering the increase in population and, especially, the erection of homes on the waste land ranging in distance from three fourths of a mile to three and a half miles to the south and southwest of the plant, we must reach the obvious conclusion that we cannot continue to go on as we are for a great many more years, especially if a firm supply of ore concentrates become available."[97]

FIGURE 2.1 Tacoma smelter and surrounding community, 1953.
Courtesy of Tacoma Public Library, D76310–2.

But with no level of government compelling the company to control its pollution in Tacoma, the cost of adding pollution controls versus the cost of business as usual appeared to be the deciding factor for ASARCO. Even in the early 1960s, the Tacoma smelter's future—most likely because of its tidewater location, far from mines supplying raw ore—looked bleak to management. Therefore, the available evidence suggests that greater sulfur dioxide recovery and additional toxic metals control was not pursued at this time.

ASARCO was accustomed to paying individual damage claims at Tacoma and at its other smelters, a solution that was less expensive than adding control technology. Near its lead smelter in East Helena, Montana, ASARCO paid to replace horses that became lead poisoned from grazing nearby. In the vicinity of the Tacoma smelter, which was growing increasingly urban by the mid-1960s, animal injuries and deaths typically concerned house pets, which were less expensive to resolve than those of livestock. The resolution of one complaint, for example, involved a thirty-dollar charge to "destroy" two dogs and ten dollars to "replace" them.[98] In comparison to multi-million-dollar investments in pollution controls, paying nominal amounts for damage to plants, lawns, and pets was much less expensive over the long run.

But individual compensation was no longer an acceptable remedy to a public that did not want to be confronted by the industrial pollution they feared was harming their health in their homes, yards, and communities. By 1962 the Citizens' Committee found that its entreaties to the local and state government had not produced tangible improvements in air quality. Next the committee appealed to the U.S. surgeon general, Dr. Luther Terry. City Manager Rowlands supported this request and made a frank appeal of his own to the surgeon general for federal help in February of 1962: "The City of Tacoma, Washington has real and serious air pollution problems caused by the operation of a copper smelter located adjacent to the City, which discharges process wastes that contain sulfur dioxide and arsenic into the ambient atmosphere. Our investigations clearly show that the sulfur-dioxide in the smelter-produced air pollution is the cause of extensive physical discomfort, widespread vegetation damage, and some other property damage in large areas within our city limits and in large adjacent areas."[99]

Reflecting the difficult political position of local officials with respect to industry within the local area, Rowlands wrote, "I am obliged to seek the assistance of an agency with higher qualifications and greater authority with respect to such matters."[100] The surgeon general was asked to conduct a formal investigation that would result in recommendations for solving the problem. In May of 1962 three USPHS air pollution experts visited Tacoma to conduct a preliminary investigation.[101]

The investigation relied on meeting with the Citizens' Committee, city and state officials, and ASARCO managers. Although the city wanted USPHS researchers to determine if the smelter was causing serious health problems as the Citizens' Committee alleged, the USPHS did not provide definitive answers. The USPHS did not collect any new data; rather, it relied on the scant data that were readily available and downplayed the health effects of the sulfur dioxide pollution in the community. The press reported that a USPHS doctor stated: "The sulphur dioxide itself probably isn't harmful in the quantities which reach the ground—but . . . it is conceivable that in a relationship with certain particles or other substances it could be harmful."[102]

A month later the USPHS wrote to the city with its recommendations and made clear that the federal government would not take a leadership role in resolving the dispute. The USPHS recommended a "long-range, area-wide program of air resource management" for the Puget Sound region, and advised the city and state to pool their resources to evaluate the regional air pollution problem and "incorporate minimum steps necessary to adequately explore that part which may involve the smelter." Sidestepping the question of whether the smelter was damaging the community's health, the USPHS simply stated: "There seems to be little doubt that pollutants arising from the copper smelter in Ruston merit further evaluation."[103]

The slow progress in controlling both smelter pollution and the more general problem of air pollution in Tacoma was not anomalous. With the exception of California, with its notorious smog problem, most states and localities had small, underfunded air pollution control programs of limited effectiveness by the early 1960s, if they had programs at all.[104] Industry opposition and lack of funding and competent staff were some of the reasons. Frustration at the slow progress led to growing calls from citizens across the country for the federal government to provide relief from the seemingly intractable problem. The federal government did expand its role in research, technical assistance, and training to the states throughout the 1950s and early 1960s. Although some provisions were made for federal involvement in interstate pollution issues, its enforcement role remained limited because it was reluctant to antagonize industry and the states, and many in the USPHS opposed a federal regulatory role.[105]

The environmental consciousness of the country continued to grow in the 1960s. When Rachel Carson published *Silent Spring* in 1962, which linked chemical exposure to environmental contamination, damage to ecosystems, and ultimately to impacts on human health, the public appeared ready to listen to and act on Carson's warnings. A new generation of scientists influenced by the kind of ecological thinking exemplified by Carson was beginning to ask the scientific questions to determine just how contaminated the modern environment was, how industrial and agricultural chemicals get into people's bodies, and what harm at various concentrations these chemicals could cause. These researchers in public health, medicine, and many other fields were in essence asking, what subtle harms is our increasingly industrialized and technological environment causing?

Although change was slow, its winds were even evident to ASARCO management by the mid-1960s, as a communication between L. V. Olson and the ASARCO's East Helena smelter manager demonstrates, referring to the work of Claire Patterson, whose research on lead in the environment exemplified this new ecological perspective. Olson wrote, "One author recently published a ridiculous article in *Archives of Environmental Health* on the dangers of lead in our present civilization; he even has measured a marked increase of lead in the ocean water! I am sure such statements can be successfully challenged, but with the present day developments, we must always be prepared."[106]

Environmentalism as a social movement gained momentum in the 1960s. The decade saw the political reinvigoration of the Sierra Club as a force for environmental improvement as well as the development of organizations such as the National Resources Defense Council. Greenpeace would form in the early 1970s, and hundreds of local organizations and chapters of national groups would form across the country. Influenced by other social movements of the time, such as the civil rights movement, the environmental movement was challenging industry's right to pollute and pointedly questioning existing power relationships in American life. Environmentalists were increasingly defining a healthy environment as a fundamental right.[107]

During the summer of 1966 warm, stable air once again trapped smelter fumes over the city, and complaints from residents poured in.[108] ASARCO even made what it called "reconnaissance flights" over the city to study the extent of its pollution plume. In a confidential memo, Shinkoskey told Kenneth Nelson, the company's new director of the Department of Hygiene and Agricultural Research, that his aerial observations were "anything but encouraging. . . . From on top the smoke was visible for about three miles out from the stack in a fan which was two to three miles wide at its farthest point from the stack." In fact, that summer's pollution problems prompted Shinkoskey to recommend that the company study the possibility of once again increasing the height of the stack—this time to 813 feet.[109]

By that fall the city council was taking on the smelter's pollution problems with a vigor not seen before. One councilmember said that air pollution was the most frequent citizen complaint that council members received. Several members called for immediate action and focused blame squarely on the smelter. One who had his own property damaged that summer said: "I think we have to reach the point where finance takes second place. We have to decide if we still should keep these industries here if they're impairing the health of our citizens." Other council members voiced the ongoing concern that industry would relocate if forced to clean up.[110]

But that year a budget shortfall prevented significant new funding for the problem, despite the protests of citizen groups including the League of Women Voters, the Pierce County Medical Society, and the Citizens Committee on Air Pollution.[111] Coinciding with stalled progress in Tacoma, the state legislature was considering several air pollution control bills that would put enforcement authority in the hands of the state, and it passed a state Clean Air Act in 1968.[112] The state's action seems to have emboldened Tacoma's leaders who, coming up on another summer smoke season, once again took up the issue of smelter pollution in May of 1967 and appropriated $10,000 to sample sulfur dioxide concentrations in the West End. That spring the city also brushed aside the forty-year-old argument that it could not act against the smelter because it was outside of the city's jurisdiction, passing an ordinance to set limits on ground-level and stack concentrations of sulfur dioxide.[113]

Of the ordinance, city attorney Marshall McCormick said: "It just seems to me it is common sense that you should be able to protect your city from nuisances and danger—no matter where they may originate."[114] City Manager Rowlands said that smelter pollution was an "emergency" and urged ASARCO to take action. In a letter to ASARCO he wrote: "Unless this situation is corrected in the immediate future, it is quite probable that the city of Tacoma will request the governor to ask the secretary of health, education and welfare and the state air pollution control agency to carry out a large-scale investigation of this intolerable condition."[115]

For its part, ASARCO did not capitulate to the new demands. Rather, it argued that that smelter's health effects were overstated. To rebut growing concerns about health effects, Dr. Sherman Pinto, ASARCO's medical director, met with the director of the Tacoma-Pierce County Health Department to share with him the findings from studies of workers at the Tacoma smelter, which apparently showed that sulfur dioxide was not related to their mortality.[116] The argument went that if workers were not dying from exposure to sulfur dioxide, then there was little reason to be concerned about health effects in the community as concentrations were invariably lower.

It is clear that citizen concerns, complaints, and organizing empowered city officials to finally take action. Without the push from community members, it is unlikely that Tacoma's leadership would have reversed itself and decided that it could, after all, pass an ordinance aimed at the smelter. Passage was likely due a realization on the city's part that times had changed. Although residents had protested smelter pollution all century long, their protests became more organized and reached a crescendo by the late 1960s. Residents simply would not tolerate pollution of the environment at the levels that were occurring. There was growing and convincing scientific evidence that community health could be harmed by air pollution. The idea that pollution was simply a nuisance had been effectively overturned. These changes, of course, had occurred not just in Tacoma, but across the United States during this time period, aided and abetted by government investment in independent scientific research in public and environmental health.

City leaders may also have been emboldened by the state's Clean Air Act, a show of the state's political will to control air pollution. Statewide legislation as well as looming federal standards would limit industry's ability to threaten to relocate to find more lax air quality standards. A regional air pollution control authority was being formed, the Puget Sound Air Pollution Control Agency (PSAPCA), which would almost certainly take steps to curtail the smelter's pollution. The city would be able to point to the ordinance as proof that it had done something about the smelter.

Tacoma's sulfur dioxide ordinance, passed in July of 1967, limited only ground-level concentrations of sulfur dioxide.[117] It did not require ASARCO to further control or capture sulfur dioxide; dispersion would still be allowed. The issue of how sulfur dioxide pollution would be reduced was about to become a hotly contested issue in smelting states after the passage of the federal Clean Air Act of 1970. Dispersion and meteorological curtailment, presented as a temporary solution to the problem in the 1920s, would still be ASARCO's and some other companies' preferred solution at the end of the 1960s.

On the threshold of the 1970s ASARCO continued to resist pressure to solve the Tacoma smelter's pollution problems. The smelter's capacity for metals recovery had not been significantly upgraded since the 1930s, and little was being done to reduce sulfur dioxide emissions.[118] Although the company denied that smelter operations caused important health effects, there was a growing body of scientific evidence regarding the health effects of sulfur dioxide, lead, and arsenic.

But Washington State finally seemed resolved to seriously tackle air pollution. The newly formed PSAPCA was about to make the Tacoma smelter its number one priority. Even in this new context, in which state and local governments were gaining power to act against polluters and the federal government's authority was being concretely defined, ASARCO still managed to assert some control over the dialogue about smelter pollution and its impacts by denying serious health effects from smelter pollution, threatening to close if forced to make expensive upgrades, and citing industry-sponsored science to support its views.[119] These strategies remained potent tools even in this new environment.

In the next few years, the discovery of a lead poisoning epidemic near ASARCO's smelter in El Paso would thrust the company's pollution into the national spotlight. El Paso was the first smelting community in the United States in which widespread heavy metal poisoning was discovered in children living nearby. El Paso was a public health crisis, one that threatened the health of thousands of children.

3

Uncovering a Crisis in El Paso

In the southwest border town of El Paso, Texas, ASARCO operated another massive smelter for much of the twentieth century. In contrast to Tacoma's location on the foggy and rainy shores of Puget Sound, the El Paso smelter rose out of the parched Chihuahuan Desert, producing lead and copper from raw ore brought by rail from Mexican mines. In El Paso, the smelter's fires were stoked largely by Mexican American workers who raised their children in the shadow of the stack.

A prominent El Paso company since the turn of the century, ASARCO was apparently surprised to learn in the early 1970s that a previously "friendly" city administration was scrutinizing the smelter's contribution to air pollution and growing increasingly concerned about its impact on the health of people who lived in the city.[1] That El Paso's political leaders shared the emboldened attitude of Tacoma's toward ASARCO in the late 1960s was not a coincidence; rather, it had to do with the growing national environmental consciousness and mounting concerns about the health effects of air pollution. These local efforts were buoyed, and in some cases made possible, by pending action at the federal level. Specifically, the federal government was about to gain new powers to control air pollution, with the passage of the 1970 Clean Air Act (CAA), which provided for federal minimum standards and enforcement. The CAA was signed into law by President Richard Nixon in the same year that the Environmental Protection Agency (EPA) was established and the federal government's regulatory and research interest in occupational safety and health was codified in the Occupational Safety and Health Administration (OSHA) and its research arm, the National Institute for Occupational Safety and Health (NIOSH).[2]

One of the first issues that the EPA had to tackle was sulfur dioxide emissions from heavy industry; smelters were a significant contributor to this issue, the most significant source of sulfur dioxide emissions in the West, second only to electric utilities nationwide.³ In 1969 sulfur dioxide emissions from the country's nonferrous smelters were quantified for the first time in a federally sponsored study. The study found that nonferrous smelters were pumping 1.9 million tons of sulfur oxides into the air each year; of those smelters, copper smelters contributed the most: 1.4 million tons. Notably, the problem was concentrated west of the Mississippi River, where 97.4 percent of all emissions occurred. On the threshold of the 1970s U.S. copper smelters were controlling only about 19 percent of the sulfur dioxide generated, which reflects the long-held preference in the industry for dispersing, rather than capturing the toxic gas.⁴

The smelting industry's response to imminent local, state, and federal controls on sulfur dioxide emissions was a harbinger of how it would handle other challenges to a polluting status quo. Copper companies in particular were organized in their opposition to strict standards and tried to influence the EPA to weaken a proposed federal standard to require 90 percent control of their sulfur dioxide emissions.⁵ The EPA did not promulgate the 90 percent standard for copper smelters, even though it was thought to be technologically feasible; instead the federal standard would require about 51 percent control. The EPA also urged states to consider the costs to industry of requiring stricter standards than the federal government's, undercutting states such as Washington and Montana, which in the early 1970s wanted to take a hard line against sulfur dioxide pollution.⁶

In El Paso, scrutiny of that smelter's emissions by local officials first focused on sulfur dioxide. In the early 1970s the city and the state of Texas launched a lawsuit against ASARCO aimed at bringing relief to city residents from the smelter's sulfur dioxide pollution. It was a surprise to public officials to learn during legal discovery that that between 1969 and 1971 more than 1,100 tons of lead, 560 tons of zinc, 12 tons of cadmium, and more than 1 ton of arsenic were emitted from the smelter's stack.⁷

The discovery of the magnitude of El Paso's heavy-metal emissions would make industry's contention that toxic metals were not a threat to the health of people living near them much less convincing and would be the catalyst for official investigations of smelter pollution and community health in Tacoma and other smelting communities across the country.

From a public health standpoint the early 1970s was a pivotal time where smelters were concerned for industry, government, and communities. That the smelting industry's ability to largely control the dialogue about the harms of its emissions had been eroding over time became glaringly evident in 1970s El Paso, Texas; Kellogg, Idaho; and Tacoma, Washington, as local and federal governments took a more active role in documenting public health issues and communicating about them with the public. Indicative of changing power relationships among government, industry, and communities, local leaders in El Paso were willing to instigate a lawsuit against ASARCO, an economic powerhouse. In Tacoma, some

PSAPCA staff were said to be "brandish[ing] criminal statutes and suggest[ing] that jail terms are possible for ASARCO executives if pollution continues unabated."[8] This was the new political environment of the early 1970s, where a public that wanted the nation's air and water cleaned up was challenging industry power, and a growing government regulatory apparatus backed up public demands.

El Paso was the site of the first recognized U.S. lead poisoning disaster caused by a smelter, and it received widespread publicity. Childhood lead poisoning near the smelter had probably existed for as long as children lived nearby; however, the new political and scientific context of the early 1970s made its discovery possible. The discoveries at El Paso provoked a strong response from the affected community and from the public health researchers involved. The El Paso incident illustrates how ASARCO and other smelting industry players were responding to these new challenges from government, communities, and independent scientists. An industry research group, the International Lead Zinc Research Organization (ILZRO), played a central role, casting doubt on government research that indicated harm to children from smelter emissions. ILZRO and the lead industry took a strong interest in El Paso, which was central to an evolving national debate over subclinical damage from lead and the health threat to children from lead in air.

ASARCO's smelter in El Paso had a smokestack that was even taller than Tacoma's was. The recently demolished red-and-white striped stack stood for

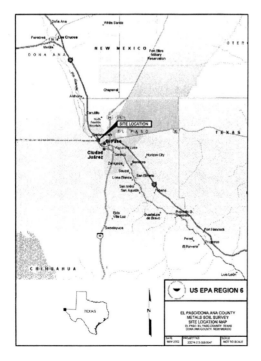

FIGURE 3.1 Map of ASARCO El Paso site and surrounding communities.

Environmental Protection Agency.

decades as a local landmark presiding over the border between El Paso and Ciudad Juarez, Mexico. At approximately 830 feet, it was a towering symbol to some of economic power and jobs, and to others, including some former ASARCO workers, of pollution, environmental degradation, and illness. No ore has been smelted at El Paso since 1999, and the plant is being demolished.[9]

When the El Paso smelter was operating, its tall stack spread pollution far and wide. Although the footprint has not been completely mapped, there is heavy-metal contamination in the cities of El Paso and Anapra, New Mexico, and on the Mexico side of the border in Ciudad Juarez that has been attributed to the smelter.[10]

Like at Tacoma, much of the ore smelted at El Paso came from foreign mines. Railroad cars brought ore from Mexico, mined from the lead deposits of Santa Eulalia, Chihuahua. To feed the copper smelter, ore was brought from Arizona. Over its operating lifetime, El Paso produced copper, lead, cadmium, zinc, and antimony. In the early 1970s the smelter employed about fifteen hundred people and was economically important to the region.[11]

The revelation that every two years the massive stack was pumping out over fifteen hundred tons of toxic heavy metals, most of it lead, alarmed health officials and touched off the first systematic study of smelter emissions and human exposure to toxic metals. El Paso would become a battleground for the lead industry because the case highlighted two critical regulatory issues: the significance of airborne lead, and the damage to health from levels of lead that did not cause clinical symptoms. Both were under intense debate at the time, and the widespread exposure in El Paso to airborne lead made it an important case for both industry and public health advocates.

As concerns about lead exposure from the El Paso smelter grew, ASARCO would turn to ILZRO, a lead industry–backed research organization closely affiliated with the Lead Industries Association (LIA). The LIA and ILZRO played central roles in funding research on lead, promoting its safety, downplaying concerns about health effects, and attempting to stave off regulation.[12] Established in 1928 by the presidents of St. Joseph Lead, ASARCO, and the National Lead Company, the LIA brought together mining, smelting, refining, and manufacturing companies to advocate, lobby, and promote the use of lead, and to conduct research on its health effects.[13] ILZRO was funded by its member corporations, mainly producers of lead, zinc, and cadmium in the United States and around the world, to conduct research on these metals. Additionally, according to Jerome Cole, ILZRO's vice president, the organization "manage[d] the environmental health affairs of the Lead Industries Association."[14]

ILZRO would play a key role in downplaying the impact of heavy-metal emissions from smelters on public health. Its research and public relations efforts were central to industry's effort to cast doubt on government research demonstrating that children in both El Paso and Bunker Hill had been harmed by smelter lead emissions.[15]

The city's lawsuit against ASARCO in El Paso began over sulfur dioxide, but it quickly became focused on the smelter's lead emissions. Lead concentrations in El Paso's air were high; in fact, in 1971 the mean annual ambient lead concentration at one measuring station near the smelter was 92.0 µg/m^3—more than six hundred times higher than today's ambient air standard.[16]

What appeared at the outset to be a local problem rapidly became a problem of national significance to public health advocates and industry alike. The investigation into childhood lead exposure in El Paso would break new scientific ground, demonstrating that children living nearby had elevated blood lead levels directly attributable to the smelter's emissions. In other words, the lead that poured out of the smelter's stack contaminated the air and found its way into children's bodies. This finding would lend support to growing calls in the early 1970s to phase lead out of gasoline and to sharply limit concentrations of lead in air. In the hopes of influencing regulatory efforts aimed at lead, ILZRO would work to cast doubt on the results of a Centers for Disease Control and Prevention (CDC) investigation in El Paso that demonstrated that children could be harmed by lead in air.[17]

Although lead had long been known to harm the health of workers, concern that lead in the environment could harm human health had been growing since the 1960s. The lead industry's control of research on its health effects, which persisted for much of the century, was being credibly confronted by this time. Researchers not reliant on lead-industry funding were challenging the industry's contention that nearly all childhood lead poisoning was caused by ingestion of lead paint.[18] The advisability of allowing lead to be added to gasoline, putting over 150,000 tons of the toxic metal into the nation's air each year, was also being sharply questioned.[19]

In 1971 lead was ubiquitous in the environment. The use of lead arsenate in agriculture for nearly a century meant that soil and food might be contaminated with both lead and arsenic. Beginning in the late nineteenth century mass production of lead paint resulted in its widespread use on the interiors and exteriors of American homes. Beginning in the 1920s tetraethyl lead was added to gasoline as an antiknock agent; during combustion this compound became airborne, filling the air with respirable particles of lead.[20]

By the early 1970s the lead industry had lost a hard-fought battle over lead in paint. No longer would significant amounts of lead be added to household paints. City governments had moved first to restrict the amount of lead that could be used in paint, and the federal government was poised to adopt legislation as well. But the market for lead paint had been dwindling anyway, as titanium oxide and other materials were increasingly replacing lead.[21]

The nation's smelters, however, were still producing lead, and the industry needed to maintain markets for their toxic product; therefore, maintaining its use in gasoline became a key goal. To do this, they needed to downplay the health concerns about airborne lead that seemed to be cropping up everywhere by the late 1960s. With the leadership of the LIA, they were determined to keep the focus on lead poisoning as a problem caused by children ingesting deteriorating paint rather than the lead-contaminated air that children all across the nation were breathing in their homes and schools and on playgrounds.[22]

Apart from the question of how lead was finding its way into children's bodies, researchers concerned with lead exposure in the early 1970s were also questioning the scientific orthodoxy regarding the level of lead in blood that could cause harm. Industry-affiliated scientists had long argued that adverse effects in workers were not seen until blood lead levels reached 80 μg/dL. For many decades, this was considered a threshold level below which harm would not occur. Independent researchers were questioning this "threshold," particularly for children, arguing that their rapidly developing nervous systems might make them particularly susceptible to lead's toxic effects.[23]

By 1970 New York and Baltimore used 60 μg/dL as the cutoff for diagnosing lead poisoning in children; Chicago set a lower standard of 50 μg/dL. In the early 1970s the U.S. surgeon general determined that children with blood lead levels of 79 μg/dL or greater constituted "medical emergencies," those with levels between 50 and 79 μg/dL should receive medical evaluation, and those with levels over 40 μg/dL had "undue lead absorption."[24]

Although there were no nationally representative data in 1970, many in the medical and public health communities feared that far too many children were being unduly exposed and poisoned. For example, New York City childhood lead screening data for part of 1970 showed a mean blood lead level (BLL) of 20 μg/dL, with 11 percent of samples higher than 50 μg/dL. Almost 6 percent of Chicago children tested in 1967–68 had BLLs 50 μg/dL or higher. Data from various locales showed that lead exposure was widespread among the nation's children.[25]

Under the Clean Air Act, the EPA was charged with regulating airborne toxics capable of harming the public's health. By 1971 the agency had publicly stated that lead in gasoline was such a toxicant.[26] Regulation would be required. But the EPA also had another stake in limiting lead in gasoline. The catalytic converter was developed in the late 1960s to render automobile emissions cleaner, but lead in gasoline interfered with its effectiveness. The EPA would require that refiners make one grade of unleaded gasoline available to consumers by 1975 for use with cars with catalytic converters.[27] Phasing out lead from the rest of the gasoline supply would require winning arguments over the health effects of lead in air.

One of the central questions regulators had to answer was the extent to which airborne lead raised blood lead levels, particularly in children. Automobile emissions were the main source. The other significant source was nonferrous smelter emissions. This federal inquiry into lead along with looming legislation that would affect the mining, smelting, and gasoline additives industries was the larger context in which the El Paso lead poisoning crisis unfolded. The stakes were high for the industry—about 20 percent of the nation's lead production wound up in gasoline in the mid-1960s.[28] When the EPA turned to the National Academy of Sciences to evaluate the state of the science on lead as an airborne pollutant in the early 1970s, these issues were under intense debate.

In the early 1970s Philip Landrigan, a medical doctor, was beginning his career in the U.S. Public Health Service when he was asked to investigate whether children were being harmed by ASARCO's considerable lead emissions in El Paso. Landrigan was a member of the Epidemiologic Intelligence Service, a highly trained group of epidemiologists from the federal CDC, who track down sources of disease outbreaks. He was familiar with childhood lead poisoning, having seen stricken children in Boston, where he practiced prior to his CDC appointment. Landrigan was on hand in 1971 when El Paso's health director, Dr. Bernard Rosenblum, called to see if federal epidemiologists and doctors could help the city investigate.[29]

Landrigan's entry into the debate over lead exposure in El Paso, and subsequently in other smelting communities, provided critical data. His research would challenge prevailing ideas about lead exposure, such as (1) airborne lead was not an important source of exposure for children; (2) people living near smelters were not harmed by lead (and other heavy-metal) emissions; and (3) children who were damaged by lead would exhibit clinical symptoms. While Landrigan's work would be well received by the scientific community and published in top medical and public health journals, it would earn him the antipathy of some lead and smelting industry leaders.

The National Academy of Sciences' 1972 report was intended to represent state-of-the-art science on the problem of airborne lead. The committee and the consultants and contributors who produced it, however, counted a number of industry and industry-friendly scientists as members. The chair, Paul Hammond, has been described as "closely associated with the lead industry's policy positions."[30] As Landrigan's El Paso investigation was taking shape, the report by the august academy downplayed the significance of airborne lead, reporting that childhood lead poisoning "is believed to be due almost entirely to the repetitive eating of leaded house paint," and largely dismissed the public health significance of lead in smelter emissions even though animal deaths near smelters were known to occur.[31] The report both reflected the scant scientific attention given up until this time to addressing community health effects from heavy metals in smelter emissions and reinforced industry's long-held position that smelter emissions did not harm health. The contention, of course, also supported the larger industry goal to downplay the importance of airborne lead at the very moment the EPA was evaluating it in the context of setting federal standards for air pollutants. The El Paso investigation was about to demonstrate that the National Academy committee could not have been more wrong about smelters, their lead emissions, and impacts on human health.

Dr. Bertram Carnow suspected that the El Paso smelter's lead emissions could be harming children living nearby, and he was apparently the first to test them. Carnow was a medical doctor, a professor at the University of Illinois School of Public Health in Chicago, and a specialist in occupational and environmental health.[32] Hired by the City of El Paso to help prepare for the lawsuit against ASARCO, Carnow learned of the smelter's lead emissions in 1971 and recommended testing the children who lived closest to the smelter to see if their blood lead levels were elevated. A number of children who lived in Smeltertown, a low-income Mexican American community adjacent to the smelter gates, were tested and found to have elevated blood lead levels. It was this finding that prompted Bernard Rosenblum to contact the CDC to ask for help.[33]

Rosenblum had several questions that he needed CDC's help answering: Was the smelter the cause of elevated blood lead levels in Smeltertown children? If so, how geographically widespread was the problem? Were children being harmed? In the early 1970s the CDC possessed limited capacity and expertise in environmental health issues in general, and little expertise specific to lead. The CDC had been established in 1946 with a focus on infectious disease control, originally mosquito vector control for malaria. Over time the CDC developed a strong focus on vaccine preventable illness. By the 1970s society's concerns about environmental influences on health were pushing CDC scientists to broaden their purview.[34] But when Rosenblum contacted the CDC in 1971, linking human health problems to environmental contamination was a field in its infancy, and one with which CDC had little experience.

Meanwhile, in El Paso as in Tacoma, ASARCO generally enjoyed congenial relations with the city as well as strong support from the local chamber of commerce. Therefore, the city's lawsuit caught ASARCO off guard since it was initiated by what was regarded as a "friendly" city administration. The company was also surprised to learn during a pretrial deposition that Smeltertown children had elevated blood lead levels. It began its own testing of children in Smeltertown and by mid-trial had referred seventy-two children with blood lead levels ranging from 40 to 82 μg/dL to Dr. James L. McNeil, a local pediatrician. The highest blood lead level measured was 94 μg/dL.[35]

Landrigan's investigation began in the winter of 1971–1972, as the city's case was nearing trial. Although Smeltertown, because of its proximity to

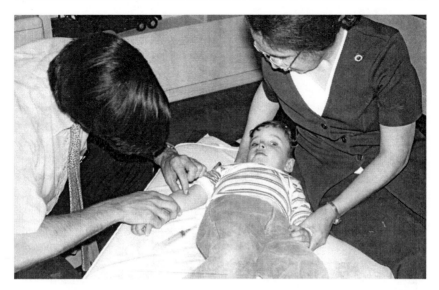

FIGURE 3.2 Dr. Stephen Gehlbach of the Centers for Disease Control conducting blood lead testing in El Paso, 1972.

Courtesy of *El Paso Times*.

the smelter, was likely the epicenter of the problem, Landrigan strategically began testing children living in the El Paso neighborhood of Kern Place, then a more affluent, predominantly white community about a mile northeast of the smelter.[36] Had he started in Smeltertown, findings of lead poisoning in lower-income Mexican American children might have been dismissed as the fault of poor, uneducated parents who did not responsibly tend to their children. The lead industry had used arguments like this, playing on racial and class stereotypes, for many decades to shift blame for lead's damage.[37] If Landrigan found that affluent children had elevated blood lead levels, this meritless argument could be easily dismissed.

Working with his colleague, Dr. Stephen Gehlbach, Landrigan reviewed air sampling data and other records, and from the available preliminary data, he recalled, "It was clear there was a problem [with environmental lead]. We weren't sure if there was a human health problem or not."[38] The tests of Kern Place nursery school students turned up blood lead levels above 40 μg/dL in two-thirds of those tested.[39] Further testing in early 1972 found that 62 percent of children ages ten and under, and 43 percent of people of all ages within a mile of the smelter had BLLs equal to or greater than 40 μg/dL.[40] *Time* magazine reported that some exposed children had basophilic stippling, lead lines in their gums, and foot drop, all conditions that are attributable to lead poisoning.[41]

Because of its proximity to the stack, Smeltertown, was not surprisingly one of the hardest hit areas. By March of 1972 five Smeltertown children had been hospitalized and thirty more were said to be suffering from chronic lead poisoning.[42] El Paso's mayor, Bert Williams, sought federal assistance for the situation, described as "one of the nation's first lead poison disaster areas."[43]

In an attempt to discern how geographically widespread the problem was, the CDC next took blood samples from a random sample of the population within four miles of the smelter. By August of 1972 the CDC estimated that 2,700 residents between the ages of one and nineteen in south and west El Paso had BLLs of ≥ 40 μg/dL.[44]

The widespread public health impact of the smelter was documented through a well-designed epidemiologic study that challenged ASARCO's contention that the problem was a localized one largely confined to Smeltertown and largely attributable to exposure to contaminated soil.[45] Notably, the Lead Surveillance Committee of the El Paso County Medical Society did not support widespread blood lead sampling to identify the extent of the public health problem. The committee's minutes from the summer of 1972 state that any "further massive blood lead sampling outside of the Smeltertown–Old Fort Bliss area is at this time unjustified, based on the data available to us."[46]

Carnow also conducted his own research in the community, which included medical evaluations of ten children with elevated blood lead levels. Of the ten children he examined, "four had moderately severe anemias, three had abnormal electroencephalograms, one had a foot drop, two appeared to be retarded, and an additional number gave a history of progressive difficulty in carrying out their school work. Four gave a history of appetite loss, lassitude, abnormal fatigue, and frequent abdominal complaints."[47] The symptoms were consistent with lead exposure and poisoning.

When the crisis hit the front pages of major newspapers across the United States in March of 1972, ASARCO was urged by its local attorneys to close the El Paso plant, which it apparently "seriously considered." But according to an internal industry document, ASARCO's main concern was "keeping the plant open." Further, "the prospect of liability to the 100 families involved was secondary."[48]

To defend itself from allegations that the smelter's ongoing emissions were poisoning children, ASARCO embraced the so-called soil theory. It pinned the blame for lead poisoning in Smeltertown primarily on ingestion of lead that had built up over time in soil. Lead in the air was thought to be of lesser importance. At the time of the trial ASARCO's scientists were said to be "doubtful that the ambient air levels were sufficiently high to be causing the high blood lead levels."[49] The soil theory was thought by ASARCO to have been "sufficient to keep the plant open."[50]

A finding that children's elevated blood lead levels were caused by airborne lead might have tempted regulators to shut down the smelter to address the problem. Additionally, the soil theory also had the benefit of being consistent with the lead industry's position that ingestion of lead was the most important source of exposure for children. By implication, if children's blood lead levels weren't significantly raised by the up to 92 µg/m^3 of lead in the air near the smelter, then the <5 µg/m^3 in the air in most American cities at that time, caused primarily by burning leaded gasoline, should be safe. During the trial ASARCO removed some of the topsoil in Smeltertown and added chemical binders to suppress lead dust.[51]

When the CDC published detailed results of its investigation in 1975, Landrigan and his coauthors largely blamed airborne lead—both inhaled and ingested as settled dust—for the children's elevated blood lead levels.[52] ASARCO admitted in a private meeting in 1974 "the ambient air was more important than they thought."[53]

All things considered, ASARCO fared well throughout the crisis and the trial. In fact, the company felt "'lucky' to have pulled through."[54] It did not have to close down, it was not forced to pay the $1 million the city had originally sought, nor did it have to admit liability for harming children with toxic metals. A consent decree with the city stipulated that it would "pay medical expenses for [the] diagnosis and treatment of children with high blood lead levels" as well as $80,500 in fines.[55] ASARCO agreed to the consent decree "at the urging of local counsel for the purpose of maintaining as much good will as possible."[56] In total, ASARCO shouldered approximately $350,000 in "civil penalties and court costs," and paid for medical treatment for 148 children.[57] In addition, it agreed to install a new sulfuric acid plant and agreed to a "permanent injunction against emission of heavy metals into the atmosphere," but this would not come to pass.[58]

At the time of the settlement, two El Paso papers praised ASARCO's corporate citizenship and criticized the city and the Texas Air Control Board for bringing negative publicity to the city.[59] By 1974 two complaints that included fifty plaintiffs had been filed against ASARCO. The company planned to settle both because it wanted "very much to dispose of the matter now."[60]

Smeltertown residents had complicated responses to their community being called a "lead poisoning disaster area," and to the decision to raze their homes, which can still provoke emotional responses by former residents. The people who lived there were mainly smelter workers and their families, and many had feelings of loyalty toward the company that employed them and a deep connection to the community they built in the shadow of the smelter's stack.[61] People lost homes that had been in families for generations, and the tight-knit community was dispersed throughout the city. Some residents sought to downplay the risks that their children faced since smelter pollution had long been a fact of life. They reasoned that if the health effects of elevated blood lead were serious, this would have long ago been obvious. Others suspected the motives of city officials who they thought had not previously expressed concern about their community's health and well-being. Community residents were offended by press reports depicting their community, in which they had considerable pride, as dirty and unsafe.[62] The razing of Smeltertown forced the relocation of five hundred people.[63]

ASARCO's available internal communications about the crisis do not reveal a great deal of concern about the fate of children who had elevated blood lead levels or were lead poisoned. Smeltertown appears to have been razed to appease the city, not because of concerns about ongoing health consequences for residents. Notes from a meeting with ASARCO officials describe what happened in this way: "After the consent decree the city insisted something more be done about Smeltertown. ASARCO negotiated with the owners of the land, then evicted the tenants and leveled the town."[64]

When the El Paso disaster occurred, little was known about effects of lead poisoning in children that were not clinically detectable by physicians. Typically it was only at high blood lead levels that children experienced observable symptoms. Philip Landrigan and others at the CDC realized that the El Paso disaster represented an opportunity to understand the more subtle, less readily observable effects or subclinical manifestations of lead poisoning through epidemiologic studies. This too had implications for the EPA's pending regulatory actions on lead. If lead exposure caused more subtle harms detectable at lower levels of blood lead, this would strengthen growing calls to remove lead from gasoline and to set a more stringent standard for lead in air.

To answer this question, Landrigan and colleagues planned a study that would test IQ and other neuropsychological parameters in El Paso children to examine lead's effects on their functioning. What they didn't realize was that ASARCO was planning a competing study, one that would be funded by industry, and the company had gained the support of a segment of the local medical and public health communities.

Landrigan's proposed follow-up study was discussed at a meeting of the El Paso Board of Health in the spring of 1973. He was seeking approval to conduct a CDC-backed study. Also discussed was the study proposed by ASARCO to be led by Dr. James McNeil. The ASARCO-backed study was actually funded in equal thirds by ILZRO, ASARCO, and the Ethyl Corporation (producer of tetraethyl lead, the gasoline additive).[65] The involvement of ILZRO and the Ethyl Corporation signified the broader strategic importance of the questions being examined in El Paso to the lead industry. ILZRO played a key role in coordinating the study, appointing a steering committee, providing technical assistance to Dr. McNeil, and attempting to disseminate the findings of the study widely.[66]

McNeil's proposal to ILZRO for study funding lacked the kind of detail and specificity that would be required in many competitive funding situations. McNeil also appeared to be skeptical that Smeltertown children may have been harmed by lead. For example, his protocol stated that he would conduct EEG's on the children because the tests "may yield useful negative information," and that "clinically, I have not been able to detect any evidence of damage from prolonged exposure to lead in Smeltertown despite the constant suggestion in the literature that such levels on a prolonged basis may be damaging."[67] It was also clear that the lead industry would control the study because ILZRO would appoint a "special advisory committee . . . made up of representatives from the lead industry and others selected for their specialized knowledge in the fields required by the study."[68] The steering committee included researchers such as Donald Barltrop, Julian Chisolm, and Henrietta Sachs, all of whom received funding from ILZRO during their careers. Both Chisolm and Sachs viewed exposure to lead paint rather than airborne lead from gasoline combustion as the more important source of children's exposure to lead in the environment.[69] Also on the committee were representatives of the Ethyl Corporation, the LIA, and DuPont.[70] The agreement between McNeil and ILZRO for the $100,000 study stipulated that manuscripts would be submitted to ILZRO for "criticisms and suggestions" prior to publication. Also McNeil would not be able to disclose ILZRO as the study funder without its prior consent.[71]

According to Rosenblum, McNeil told the El Paso Board of Health that his study was "identical" to the CDC study and would be a better deal for taxpayers as it was being privately funded. The board decided to officially endorse the McNeil/ILZRO/ASARCO study. Rosenblum wrote to Landrigan and told him the bad news: "the Board unanimously voted to cancel the remaining portion of your study and in its place accept Dr. Mc Neil's study from the International Lead Zinc Organization."[72]

CDC investigators enter states and localities by invitation of the respective officials only. Thus, they are dependent on good relationships and political willingness for the federal government to be involved. Notwithstanding the cooperation Landrigan initially enjoyed in El Paso, he found himself barred from conducting further research in the spring of 1973. Undeterred, Landrigan appealed directly to the Texas attorney general who, as Landrigan put it, "made phone calls." A few days later the CDC investigation was back on. Landrigan recalled, "He invoked some kind of political power."[73]

Landrigan's follow-up study, which began in the summer of 1973, compared children with blood lead levels above 40 μg/dL to those with lower levels on measures of neuropsychological functioning. This research would lend credence to developing concerns that subclinical lead exposure could impair children's nonverbal cognitive and perceptual motor skills.[74]

The ILZRO study was also conducted in 1973. It was intended to examine the "subtle long-term effects" of lead and was a key element in ASARCO's response to the crisis as well as the lead industry's response to the burgeoning science of subclinical lead poisoning and airborne lead exposure. McNeil had an ongoing relationship with ASARCO as the pediatrician who had been called upon by the company to test blood lead levels in Smeltertown children before and after the trial. He was also responsible for the care of many of the Smeltertown children who were found to have high blood lead levels.[75]

McNeil's study compared children who had formerly lived in Smeltertown to two control groups—El Paso children living outside of Smeltertown and a rural control group. Landrigan alleged that the Smeltertown group included children who had lived away from the smelter for at least a year as well as those who had received chelation treatment.[76]

McNeil's findings were to be publicly released at a meeting of the World Health Organization in Paris in June of 1974. In preparation for this, ILZRO's public relations firm, Hill & Knowlton, the firm responsible for the tobacco industry's campaign to refute the link between smoking and lung cancer during the 1950s and 1960s, was working to ensure that the study received favorable treatment in the press in the United States. The news release drafted by Hill & Knowlton was titled "El Paso Study Finds No Subtle Damage to Children Exposed to High Lead Levels."[77] Hill & Knowlton thought it would be best if Texas Tech University, McNeil's institution, "would agree to be the source of the release . . . since it would, in effect, provide a third party endorsement."[78] The public relations firm also sought to have the study released widely to "key decision makers on the federal level," including EPA leadership, the White House domestic counselor, congressional committees, the Interior and Commerce Departments, and the Office of Management and Budget. Attesting to the importance of El Paso to the question of lead in gasoline, Hill & Knowlton suggested to Philip E. Robinson, executive vice president of the LIA, that "a special letter ought to go to Senator Biden who headed the Senate panel on lead in gasoline regulations. It could help to reinforce some of the doubts he expressed at the hearing."[79]

McNeil's presentation to the Paris meeting concluded that "children who are healthy, well nourished, and not anemic may carry significant elevations of blood lead levels in the range of 40 mcg percent to 80 mcg percent (40 to 80 μg/dL) over a period of years without apparent deleterious effects."[80] McNeil formed this conclusion despite not having fully analyzed his data.[81]

In July of 1974 ILZRO was pushing McNeil to finish the study and prepare a manuscript, and was reaching out to researchers in the lead industry to find an expert in pediatric psychological evaluation.[82] McNeil also needed significant help with the statistical analysis. A biostatistical consultant to ILZRO, Philip Enterline, raised questions about the validity of some of the data and pointed out that McNeil may have used the wrong statistical test in his preliminary analysis.[83] McNeil believed that Smeltertown children had several characteristics that accounted for his findings, despite their blood lead levels. The children had good nutrition, were not anemic, and were exposed to lead, he thought, largely through soil. All of these factors, he believed, resulted in a relatively low risk of adverse effects from lead exposure. He contrasted the Smeltertown children with urban children who experienced adverse effects from lead exposure "with the lower blood lead value that is compounded by paint, pica, anemia, and malnutrition."[84]

Throughout the process of conducting the study, analyzing it, and interpreting it, McNeil collaborated closely with Jerome Cole of ILZRO. Cole provided financial support, encouragement, and advocacy on behalf of the study.[85] But McNeil had difficulty finding an academic journal that would publish the study. It was rejected by the *New England Journal of Medicine*, and the *Archives of Environmental Health* asked him to rewrite it before it could be published.[86] One of the criticisms of the study was that it had selected exposed children based on their prior residence in Smeltertown rather than on their blood lead levels, therefore children could not be strictly categorized as exposed or unexposed on the basis of their blood lead levels.[87] Landrigan publicly criticized the study's methodology, and one member of the study's steering committee resigned in protest, calling the study a "whitewash" and asking that her name not be associated with it.[88]

By November of 1977 McNeil's El Paso study still had not been published in a peer-reviewed journal. Probably because the study was so important to the lead industry, Cole offered ILZRO's ghostwriting services to McNeil. Cole told McNeil, "We would like to have a crack at preparing a publishable article based on your work." Then McNeil would submit the article under his name to a journal.[89]

Despite the difficulty ILZRO and McNeil had publishing the study, it was still used to cast doubt on Landrigan's findings. The findings were also used successfully to downplay health concerns in El Paso, likely saving ASARCO legal costs, and to support industry goals regarding the regulation of lead in air.[90] ILZRO, in particular, used the study to advance its aim of forestalling regulation of lead in air, and it paid for McNeil to present his findings at an EPA air standards hearing on lead and at a similar California hearing.[91]

Another interesting twist to the controversy that speaks to the importance of the studies beyond El Paso was that a committee convened by the Society for Occupational and Environmental Health and headed by Warren R. Muir of the White House Council on Environmental Quality was appointed to resolve the controversy around the two studies. A summary of the committee's findings was included as an addendum to the EPA's 1977 criteria document on lead. The committee found that both the McNeil and Landrigan studies had methodological flaws; therefore, "no firm conclusions could be drawn from the studies as to whether or not there are subclinical effects of lead on children in El Paso."[92] Not mentioned was the rigorous vetting process that Landrigan's study had been through before publication in the *Lancet*.

Landrigan's methodical research, his refusal to downplay his findings, and the undeniable existence of lead exposed and lead poisoned children across El Paso made him a threat to the industry. In an internal industry document, ILZRO deputy director Jerome Cole was quoted as describing Landrigan as someone who "wanted to make a name for himself in El Paso." Cole apparently felt that ILZRO had "cut him off."[93] For many more years, where lead and smelters were concerned, Landrigan would earn industry's ire.

ASARCO never ceased emitting lead and other toxic metals into El Paso's air until it closed the smelter in 1999. In 1977 lead emissions from the smelter had decreased, but they were not completely abated, and El Paso's air was reported to still have some of the highest concentrations of lead and other heavy metals in the United States. Children continued to be exposed to air emissions as well as lead that had built up in soil. Mexican health officials estimated in 1974 that more than eight thousand children in Mexico under the age of ten were "suffering because of lead pollution from the smelter."[94] By 1990 the smelter was still putting nearly 100 tons of lead and 29 tons of arsenic into the air annually. And a study published in 1997 found that among children living in the area of Anapra, Ciudad Juarez, closest to the U.S. border, 43 percent had BLLs >10 µg/dL. In 2006, raising new concerns for El Paso residents, ASARCO was accused of illegally burning hazardous waste at the smelter in the guise of recovering metals from hazardous waste. More than 5,000 tons were allegedly burned, including 300 tons of "nonmetallic waste" from the Rocky Mountain arsenal chemical warfare depot.[95]

El Paso was the opening salvo in what would become an all-out fight over community health near key nonferrous smelters in the West that would go on for at least another decade. The El Paso lead poisoning crisis was a threat to both the smelting and closely tied lead industry. The CDC's studies in El Paso demonstrated that airborne exposure to lead could be significant in raising blood lead levels and that children's health could be harmed at lower levels of blood lead than previously thought. The implications of both of these findings for regulation were clear and were the main reason that El Paso provoked a strong response from ASARCO, ILZRO and the lead industry.

For ASARCO, El Paso was a lesson in successful crisis management, and it provided a template for Bunker Hill as the company responded to the lead poisoning epidemic in northern Idaho, where it pursued three key aims: to continue operating, forestall regulation, and limit its liability for injuring nearby residents.

However, for community residents concerned about smelter pollution elsewhere, particularly in Tacoma, the discoveries in El Paso were emboldening and led to more concerted efforts to push local and state governments to examine local environmental contamination and community exposure. And for environmental and public health scientists as well as regulators, El Paso was a wake-up call that established a clear link between community health and industrial emissions—it furthered a growing interest in subclinical lead poisoning and set a high standard for scientific rigor in community environmental health investigations.

Importantly, for the first time the CDC's voice was added to debates over community health effects from smelters, and it quickly became a strong and credible counterpoint to industry's claims. Landrigan's research in El Paso contributed compelling evidence to the growing body of health research that would underpin federal regulation of lead in air, resulting in significant public health gains.

After El Paso, the Bunker Hill Mining Company began a secretive sampling program in the Silver Valley to test children's urine for lead. After learning of ASARCO's experience in El Paso, Bunker Hill would oppose efforts by the CDC, and Philip Landrigan in particular, to direct the response to the lead poisoning crisis in Idaho. It is telling that two of the first things Bunker Hill did in response to the crisis was to join ILZRO and to consult with ASARCO leadership to learn how ASARCO successfully navigated the El Paso crisis.[96]

4

Bunker Hill

In the spring of 1972, within a few weeks of the El Paso lead poisoning crisis becoming public, ASARCO chairman Charles F. Barber sent a letter to Frank Woodruff, the president of Bunker Hill Mining Company near Kellogg, Idaho. Barber was writing to alert Woodruff to the lead poisoning problem found in El Paso. He sent along an internal ASARCO report on the matter, which summarized ASARCO's views on and response to the crisis.[1] Woodruff oversaw the operations of the Bunker Hill smelter in northern Idaho. Built in 1916, Bunker Hill was an enormous lead-zinc smelter that processed much of the vast mineral wealth that was laboriously extracted from the deep mines of Idaho's Silver Valley. The smelter took in raw ore and turned out silver, gold, lead, mercury, cadmium, and zinc. By early 1980 the area around it had earned the dubious

FIGURE 4.1 Pollution pours out of the Bunker Hill smelter, 1969–1970.

Courtesy of MG 367, Special Collections and Archives, University of Idaho Library, Moscow, Idaho.

distinction of being the site of the "worst community lead exposure problem in the United States."[2]

Barber's note did spur Bunker Hill to act—top managers initiated an investigation into whether children living near the smelter were being poisoned—a seemingly responsible step for the company to take. However, this was not an ordinary public health investigation. Rather, it was a secretive study of local children that was also methodologically flawed.[3] Despite the study's flaws, a prudent interpretation of the results would have led to an acknowledgment that children in the Silver Valley were being exposed to lead in 1972, and some were almost certainly poisoned, particularly children who attended the elementary school closest to the smelter. Bunker Hill, however, did not interpret the study results this way and did not alert the community that children could be at risk of lead poisoning or reduce its lead emissions. Instead lead emissions drastically increased in the coming years, and a public health disaster was uncovered.

Although it may seem incomprehensible that Bunker Hill could surreptitiously study local children in the 1970s, this was possible because Kellogg, Idaho, was very much a company town where few residents or community leaders were inclined to ask critical questions of a dominant employer. The pointed questioning of industry's right to pollute that was occurring in Tacoma and elsewhere across the United States was not happening to the same degree in Idaho's Silver Valley, where the mining and smelting industries were significant players in the state's economy and politics. When the lead poisoning crisis became public in the ensuing years, the company experienced public support from critical segments of the community, including some local doctors, many in the business community, some in public health, and local and state government, even though Silver Valley's lead poisoning epidemic was arguably the worst the United States had ever seen from an industrial point source.

The Bunker Hill disaster starkly illustrates how high the cost of lead smelting can be in the absence of effective regulation for people living in nearby communities, particularly children. A detailed look at how the Bunker Hill disaster unfolded and the industry, government, and community response to it sheds light on the complicated dynamics of such disasters and on how politics, power relationships, science, and the "manufacture of doubt" over harm may shape the public's understanding of risk as well as the response of regulators.[4] In the face of limited community questioning, organizing, and advocacy, Bunker Hill largely succeeded in controlling the public discourse over risk and harm after the lead poisoning disaster was discovered. The central role of Bunker Hill and the support the company received from the State Health Department in responding to the crisis had grave consequences for community health. Despite the diligent efforts of some government scientists and regulators, the Bunker Hill disaster shows state and federal governments largely abdicating their newly codified responsibility to protect public health and the environment. The involvement of the International Lead Zinc Research Organization (ILZRO) attests to the significance of Bunker Hill and other smelting communities in the unfolding national policy debate over regulating airborne lead.

Bunker Hill's study of lead exposure in Silver Valley children was conducted mostly during June of 1972, just a few months after the El Paso crisis became national news.[5] Dr. Ronald K. Panke, a local doctor, and top Bunker Hill managers collaborated on a plan to test local children for lead exposure by measuring lead in their urine. Dr. Panke helped to obtain urine samples from local children, and Bunker Hill provided the laboratory and the technicians to do the analysis.[6] The community was aware of the disaster in El Paso, probably because it had received significant media attention. In a later legal deposition a Bunker Hill executive explained the lack of transparency around the study: "We did not want the mothers of the community to think that we thought we had the same problem [as in El Paso], because we did not think that we had the same problem."[7]

Aside from the obvious problems with the study, the measurement chosen—random or spot samples of urine to assess lead exposure—was flawed. Urine testing for lead was known to be an inaccurate method for determining the seriousness of lead exposure in children because urine lead levels do not necessarily correlate well with blood lead levels, except when blood lead is very high (over 80 µg/dL).[8] To definitively diagnose lead exposure or lead poisoning, blood testing was the clear choice, and this was widely recognized at the time.[9] Even so, if urine samples were to be used, the preferred approach was a twenty-four-hour sample so that the amount of lead excreted during a one-day period could be measured. Dr. Panke was familiar with the medical literature at the time on testing blood and urine for lead.[10] Urine samples, however, were probably easier to obtain and required less explanation to parents and children than blood samples.

Panke was a physician at the Doctor's Clinic in Kellogg (the Silver Valley's largest town). Some of the clinic's doctors played a central role in the response to the lead poisoning crisis. In the early 1970s many people in Kellogg and some surrounding towns used the Doctor's Clinic for routine medical care. The five or six physicians on the staff were the only ones practicing in Kellogg at the time, except for one radiologist. Smelter workers who were injured or became ill on the job were typically treated by the physicians at the Doctor's Clinic, as were their spouses and their children.[11] Bunker Hill allotted the Doctor's Clinic a certain amount of money each year for fees and hospitalization costs for employees and their children. If hospitalization costs were higher in a given year, doctor's fees would be reduced to make up the difference.[12]

Panke cobbled together a combination of strategies to recruit patients for urine lead testing.[13] First he obtained urine samples from the most easily accessed children—his own patients and others at Doctor's Clinic. Local school nurses were also enlisted to collect urine samples from students. Then a special "clinic" was organized and urinalysis screening was performed on Silver King Elementary school students. The Silver King Elementary School was singled out for special study because it abutted the smelter grounds. In his report on his data collection efforts, Panke reassured Bunker management that the clinic did not arouse the concern of parents: "emphasis . . . was placed on the urinalysis screen as a whole, i.e., it was not proposed as simply as a lead screening device" and there was not "a single instance of parental disapproval or concern that we were searching for a lead problem." The last group he tested were patients from the Doctor's Clinic who lived all over the valley.[14] Adding these children to the sample from geographically more distant locations would lower the overall mean lead level found.

In all, over two hundred Silver Valley children were tested for lead exposure in 1972, and many had urine lead levels of 80 μg/L or higher.[15] There was little in the literature at the time defining "normal" concentrations of lead in urine for children. A 1964 article by Robert Kehoe reported that when urine samples in workers exceeded 150 μg /L, "the individual has probably absorbed a potentially dangerous quantity of lead."[16] Kehoe, however, was describing adult workers, not developing children, who are even more susceptible to lead. Pediatric lead expert Dr. Julian Chisolm said during a deposition in the early 1980s that total twenty-four-hour urine output in a child should not exceed 80 μg/L and should probably be closer to 30 to 40 μg/L.[17]

The highest concentrations of lead were found in samples from children who attended Silver King Elementary, next to the smelter. Some children at the school had extremely high levels of lead in their urine—the highest value measured was nearly 600 μg/L.[18]

To summarize and report on his findings, Panke averaged all of urine lead values found, even though the children were of varying ages and lived at varying distances from the smelter. In a letter report on the study to Bunker management he wrote, "We do see some elevation of our average urinary lead level in the children of this valley. This average value, however, is much below the estimated high level of normal for children." Panke summarized: "It is felt by the physicians of the clinic [Doctor's Clinic], at this time and with our current data, that we do not have a lead intoxication problem in the area children."[19]

It came to light in a deposition for a lawsuit that one of the children with a urine lead concentration over 500 μg/L attended Silver King Elementary and was a Doctor's Clinic patient. In a deposition Panke admitted that he did not notify the school, nor could he "recall" whether he notified this child's parents of his high urine lead level or the parents of the other children attending Silver King who also had high urine lead levels.[20] It should have come as no surprise that tests of children at the Silver King school showed the highest elevations in urine lead due to its proximity to the smelter. Silver King teachers told a representative of the Panhandle Health District in 1974 that it was not an uncommon occurrence for the school to be "smothered by heavy, blue smoke which caused them and many children to suffer from runny eyes and coughing. At times the teachers thought the school was on fire and rushed out in the halls only to find the smelter as the source."[21]

The urgency to medically evaluate children with high levels of lead in their bodies was well known at the time. Highly exposed children have an unpredictable risk of acute lead encephalopathy, which can result in irreversible brain damage or death. The surgeon general's 1971 report on lead considered children with blood lead levels of 80 µg/dL and above medical emergencies and unequivocally emphasized the need to immediately treat such children.[22] To be certain that none of the children with elevated urinary lead levels were at risk of lead poisoning's most severe outcomes, follow-up testing of their blood would have to be performed. There is no evidence that this was done.

After this study Bunker Hill appears to have stopped looking for a lead problem in local children.[23] However, Gulf Resources & Chemical Corporation, Bunker Hill's parent company, maintained documents on the El Paso crisis, and ASARCO shared information with Bunker Hill on the health risk to children from lead, including the surgeon general's 1971 policy statement on lead poisoning and the findings of Dr. Bertram Carnow's medical evaluation of ten El Paso children, which showed grave injuries from lead exposure.[24] Carnow's report was circulated among Bunker Hill management in 1973. One manager even expressed concern that children appeared to suffer damage with blood lead levels of 30 µg/dL, beneath the official threshold at the time for undue lead absorption.[25]

The correspondence and file documents suggest that Bunker Hill/Gulf managers were aware of the damage that lead exposure could cause, and that children living near smelters were at risk of exposure to lead. In the ensuing years, however, they would make the decision to bypass a key pollution control device, which would result in dramatic increases in lead levels in Silver Valley's air.[26] After this the CDC would document a shocking lead poisoning epidemic in the children living in proximity to the smelter.

The Silver Valley lies in the Coeur d'Alene Mountains, part of the Bitterroot Range in northern Idaho. The south fork of the Coeur d'Alene River runs through the valley and empties into Lake Coeur d'Alene, a large freshwater lake near Idaho's border with Washington State. You can drive through the Silver Valley, also called the Coeur d'Alene mining district, following Interstate 90. Kellogg is the largest town, and several other smaller towns dot the valley. Mining as a way of life is largely a thing of the past in the Silver Valley, though some locals are hopeful that it may once again drive the local economy.

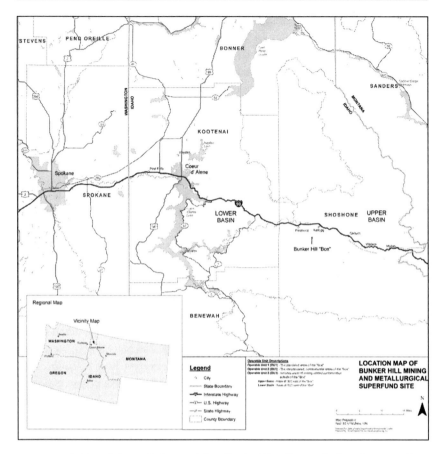

FIGURE 4.2 Map of Coeur d'Alene mining district (the Silver Valley) and surrounding areas.

Environmental Protection Agency.

Intensive mining of the valley's vast deposits of silver, lead, and zinc began with the introduction of commercial mining in the late 1880s. The mines were among the deepest and most productive in the world, following veins of silver and lead hundreds of feet into the earth. Their output over the twentieth century was immense—an estimated 17 percent of nation's lead production (or seven million metric tons), 6 percent of zinc, and 18 percent of silver. From the time that ore was first discovered in the Silver Valley until many of the mines closed in the early 1980s, it is estimated that over $26 billion worth of metals were produced.[27]

Complaints about pollution can be traced to the mining district's earliest days; historian Nicolas Casner has documented conflicts between farmers and ranchers and miners dating from the 1890s. As in other mining and smelting

FIGURE 4.3 Bunker Hill's fifty-year anniversary, July 1967. After fifty years of production, 3.3 million tons of lead had been produced.

Courtesy of MG 367, Special Collections and Archives, University of Idaho Library, Moscow, Idaho.

regions, farmers and ranchers charged that toxic mine waste discharged into the river poisoned wells, sickened and killed livestock, and destroyed their crops. In the cases that reached the courts, mine owners often prevailed. Mine owners also sought to head off such cases by buying up pollution easements and making individual settlements with farmers. When the Bunker Hill smelter began operations in 1917, more conflict ensued from smelter smoke damage, and the company bought up smoke easements to try to limit damage claims.[28]

Until 1968 mill tailings, the crushed ore left behind after being separated from that which would go the smelter, were dumped into the south fork of the Coeur d'Alene River. The tailings, which contain toxic metals, contaminated the river and migrated throughout the watershed. Today they are found in Lake Coeur d'Alene and as far away as the Spokane River, which flows out of the lake and through the city of Spokane, Washington, some seventy miles away.[29]

To say that the Silver Valley was economically dependent on mining for much of the twentieth century would be a gross understatement. Bunker Hill, one of several mining companies in the valley, employed about twenty-one hundred people, or 25 percent of Shoshone County's workforce in the early 1980s, in what were regarded as decent-paying jobs, despite tough and dangerous work.[30]

The company was called "Uncle Bunker" by many—an acknowledgment of its influence, reach, and power in this small community. Bunker Hill was an Idaho company until 1968, when it was taken over by Houston-based Gulf Resources & Chemical Corporation. Historian Katherine Aiken regards this change in ownership as marking a major shift in corporate culture. Gulf's emphasis, from the time it took over Bunker Hill and until it closed, was on squeezing as much profit as possible out of Bunker's mines and the smelter.

Houston managers put enormous pressure on Bunker Hill managers in Kellogg to produce big returns for Gulf.[31]

Probably before but definitely after the Gulf Resources takeover, "Uncle Bunker's" influence and power extended far beyond the Silver Valley and Idaho, at least during the Nixon administration, which was a critical time for pollution control decision making regarding the company. Robert H. Allen, Gulf's president, had close ties with the Nixon administration, which may have aided the company in negotiating the early years of federal involvement in pollution control.[32]

Allen was a Texas-based fundraiser for Nixon's reelection.[33] In a complex series of events, a $100,000 donation to Nixon's campaign shuttled through Mexico was eventually traced back to a Gulf Resources bank account. The money was subsequently deposited in the bank account of someone arrested in the Watergate bugging incident. The *Washington Post* linked the donation to a relaxation of federal government efforts to control the Bunker Hill smelter's pollution. According to the *Post*, Bunker Hill was at the time (1972) "under pressure by the Federal Environmental Protection Agency to correct extensive water and air pollution problems. Since then [the donation] the pressure has diminished."[34]

EPA administrator William Ruckelshaus vehemently denied that any such quid pro quo had taken place and called the *Post*'s allegations "sloppy reporting."[35] But the chairman of the House Conservation and Natural Resources Subcommittee charged that EPA headquarters repeatedly refused requests from EPA Region 10 to file a civil suit against the company for water pollution in Idaho and called Bunker Hill's water discharges "a good example of the exploitation philosophy of early America."[36] Whether EPA's oversight of Bunker Hill was influenced by the campaign donation will probably never be known. A few years after these allegations were made, however, hundreds of children in the Silver Valley were found to be dangerously lead exposed and lead poisoned.

In Idaho, mining has long been big business. Idaho's congressional delegation in Washington has vigorously supported the interests of the mining and smelting industries, and its governors have tended to be outspoken in their support of the industry. The state's constitution even gives preferential water rights to mining companies.[37]

For an outsider it can be difficult to understand the central place that mining and smelting have had to long-time residents of the Silver Valley. Mining was a way of life interwoven with the cultural and physical geography of the place where ruggedness, toughness, and independence were valued. Mining was one of the few ways to support a family in the valley, and generations of men spent their working lives in deep subterranean tunnels.[38] But the jobs came at a very high cost, to the workers, to their families, and to the environment.

Going back a century, it is almost certain that children who grew up in the Silver Valley were exposed to lead and other heavy metals in their environment. Mine waste was strewn about the valley, and over time homes, schools, and apartment buildings were built over the top. The waste was used as fill and to build highways.[39] The smelter pumped out respirable particles of heavy metals beginning in the nineteen teens, further contaminating the air, water, and soil with toxic metals. Sulfur dioxide emissions from the smelter killed vegetation, resulting in a barren and dusty landscape. The environmental destruction caused by smelter emissions was obvious. Even Bunker Hill's vice president for environmental affairs, Gene Baker, admitted in a confidential talk in 1970, "our emissions have destroyed most of the natural vegetation in the valley."[40]

One of the earliest clues to a problem with lead exposure among Silver Valley children came from a 1969 federal survey near three western smelters. Researchers found that children living near the Bunker Hill smelter had the highest concentrations of lead in their hair, by far. Mean hair lead levels in Kellogg, at 107 parts per million, were more than twice as high as levels in East Helena, Montana, another lead smelting town. Children who attended the Silver King school that was closest to the smelter had mean hair lead levels almost four times higher than children who attended other schools in the valley. The Public Health Service noted at the time that the findings were "cause for concern" and recommended further sampling of entire families, including blood and urine samples.[41] Had these recommendations been implemented, perhaps some the suffering of the Silver Valley's children in the ensuing years might have been prevented.

As in other smelting communities, in the early 1970s Idaho health and environmental officials were only beginning to understand the extent of contamination of the valley's air with toxic metals. Air monitoring by the state did not begin until 1970. Between 1971 and 1972 lead concentrations in the air up to two miles from the main stack averaged between 7 and 9 $\mu g/m^3$.[42] At the time, these concentrations were high enough to be of concern. For example, in the context of evaluating lead in gasoline, the EPA wrote in 1972 that air concentrations of lead above 2 $\mu g/m^3$ could cause blood lead levels to rise, causing elevated body burdens of lead and resulting in an "endangerment to public health."[43] In the same year, the National Academy of Sciences Panel on Lead implied that lead concentrations in the air above 2–3 $\mu g/m^3$ could cause blood lead levels to increase.[44] Other conditions in the valley also increased the seriousness of air pollution—surface temperature inversions occurred about two hundred days per year, which trapped the smelter's pollution in the air that residents breathed.[45]

The Bunker Hill smelter was similar to Tacoma's smelter in that it had not been significantly modernized over the century. At Bunker Hill, the main baghouse dated to 1920 and was by any measure outdated in 1973.[46] It had been built to recover valuable metals from the smelter smoke stream and was not up to the task of modern air pollution control. And already high lead levels in the Silver Valley's air were about to increase significantly.

Cognizant of the growing public demands for air and water pollution control in the late 1960s and early 1970s, Bunker Hill did make some improvements to control its considerable air (sulfur dioxide) and water pollution.[47] But like ASARCO, Bunker Hill would balk at stringent sulfur dioxide control. In the early 1970s the state of Idaho wanted Bunker Hill to achieve 72 percent control of its sulfur dioxide, but the EPA would require 82 percent control.[48] Throughout the 1970s Bunker Hill fought the more stringent requirements, and aimed to maintain a State Air Pollution Commission that was, in a top manager's words: "sympathetic to industry."[49] In the early 1970s a Bunker Hill attorney was quoted as saying that the company would "use every trick in the book" to fight compliance dates, emission standards, and ambient air regulations that applied to them with respect to sulfur dioxide.[50]

Like ASARCO, Bunker Hill argued in favor of tall stacks and continued dispersal of the toxic gas, and it fought against strict standards for lead in ambient air. Even after the lead poisoning crisis was uncovered, Bunker Hill worked through its congressional representatives to pressure the EPA not to enforce tougher standards on lead. In the end Bunker Hill would not meet the stricter federal standard for sulfur dioxide. Rather than significantly upgrading its ability to capture more of the gas, it built a taller stack in 1977 to improve dispersion. Eventually, like ASARCO in Tacoma, Bunker Hill would enter into a settlement agreement with the EPA in 1979 that allowed for continued operation of the plant for at least three and a half more years.[51]

Although it is clear that children living near the Bunker Hill smelter were at high risk from its lead emissions, the situation was about to become rapidly worse in the fall of 1973. In September of that year, a fire destroyed parts of the smelter's main baghouse, Bunker Hill's primary heavy metal control device that filtered toxic metals from the smoke stream. The diminished baghouse capacity made working conditions even more hazardous than normal. Even before the baghouse fire, lead exposure in the plant was harming the health of workers. A study done by NIOSH in October 1971 found that 44 percent of workers tested had elevated blood lead or cadmium levels, and 24 percent had undergone chelation or received a job transfer due to lead exposure. Before the fire, in February of 1973, OSHA issued a number of citations and put the company on notice to reduce workplace exposures. But after the baghouse fire, conditions were getting worse, not better. Compared with 1971, there was a doubling in medical treatment cases and/or transfers out of lead exposure areas among workers in 1974. Because of the deteriorating conditions in the plant, workers had to wear respirators.[52]

The baghouse fire occurred at a time of increasing prices for lead and zinc and sharply climbing silver prices, and during a time when Bunker Hill was under significant pressure to generate maximum profits for Gulf. Idaho's Department of Environmental and Community Services (DECS) air quality specialist Ian von Lindern later stated that damage to a key pollution control device should have resulted in a temporary closure or production curtailment until the problem could be fixed. Instead Bunker Hill continued to operate. In fact, in late January of 1974 George Dekan, also a DECS air quality specialist, observed after an inspection of the plant that production levels appeared to be high, taking advantage of high lead prices. The baghouse was still not operating properly, due to difficulty obtaining replacement bags.[53]

By late January of 1974 the air quality problems within the plant were becoming intolerable for workers, so Bunker Hill/Gulf managers decided they would bypass the baghouse and emit heavy-metal-laden smoke directly into the air Silver Valley residents were breathing. Predictably, lead emissions jumped dramatically, from 75,000 pounds in January 1974, to 167,000 pounds in February. March emissions were even higher, at 192,000 pounds, over 3 tons per day. And lead wasn't the only toxic metal released in large quantities during this time period. In March alone the smelter pumped out over 16,000 pounds of highly toxic cadmium, over 800 pounds of the carcinogen arsenic, and nearly 1 ton of the neurotoxin mercury. All were in the air that people living in the Silver Valley were breathing. Between 1973 and the summer of 1974, an estimated 720 tons of lead were emitted into the Silver Valley's air.[54]

Several years later, in preparation for defending a lawsuit brought by lead poisoned children, Gene Baker, Bunker's Hill's vice president for environmental affairs, described the decision to bypass the baghouse as a "normal operating decision" and said that even though "emissions increased dramatically . . . they had little or no discernible impact on the community."[55] The year 1974 was a banner one for Bunker Hill profits. The company's income before taxes was about $26 million, a significant increase over the previous year.[56]

The state was aware that something out of the ordinary was going on at the Bunker Hill smelter in the winter and spring of 1973–74. Even though the state's air monitors were located about two miles from the smelter, the monitors were picking up large increases in ambient lead levels. Average daily values of lead in the air increased from 7 μg/m^3 in 1971 to 17 μg/m^3 in 1973. Twenty-four hour average samples were as high as 75 μg/m^3. Some particulate samples taken in Kellogg were found to be 10 to 20 percent lead by weight. A few samples of house dust averaged about 7 percent lead.[57]

At the time there were no federal or state standards for lead in the air, but health officials knew these levels were far too high. The data concerned Dekan and von Lindern. They realized that average concentrations measured within two miles of the Bunker Hill trumped those found at El Paso at the same distance and could have grave implications for public health.

Dekan and other state inspectors had even visited Bunker Hill at the end of January 1974, right before the baghouse bypass, to discuss air monitoring data with the company. Gene Baker apparently said that ambient concentrations of lead in the Silver Valley were "alarming" but reassured inspectors that planned new equipment would bring lead levels down. During the inspection Dekan observed that the baghouse was not functioning properly and that one-third of it was off-line. The remaining portion of the baghouse was overloaded. Repairs were occurring, but replacement bags were back ordered.[58]

Bunker Hill was on a compliance schedule imposed by the state to bring particulate emissions, including lead, under control. But it was behind schedule. Planned controls for 1972 had been pushed to 1973. And in 1973 the company's timeline for completing controls was pushed to the summer of 1974.[59]

After the fire and bypass of the baghouse, the Taylor children who lived in Kellogg were the first to be reported to the State Health Department as victims of lead poisoning. In the fall of 1973 Mr. and Mrs. Taylor took their toddler, Clara, to the Doctor's Clinic with vague complaints such as headaches, abdominal pains, and digestive problems, which can be indicative of lead poisoning. The family lived about a mile from the smelter stack. Dr. Dahlberg, a Doctor's Clinic physician, took a blood test and found the child's blood lead level was between 40 and 50 μg/dL.[60]

In January of 1974, with Clara experiencing ongoing problems, Mrs. Taylor took Clara and her younger sibling to a pediatrician in Coeur d'Alene, about forty miles west of Kellogg in a neighboring county. When Dr. Thomas Reeds saw the children, they were suffering from cold symptoms and abdominal pains. Blood tests revealed that both children had lead levels between 60 and 64 μg/dL. X-rays showed lead lines—places where lead had replaced calcium in the children's bones—at their knees, ankles, wrists, and skulls.[61]

Lead lines are considered diagnostically definitive of lead poisoning. Even without enough exposure to produce lead lines, much of the lead that is retained in the body will be stored for decades in bones, where it can cause ongoing health problems. In men, bone lead is associated with an increased risk of death from all causes and with cardiovascular disease later in life.[62] For women, the calcium demands of pregnancy can lead to the mobilization of lead from bone, affecting both mother and baby. Bone lead in pregnant women is associated with an increased risk of maternal hypertension, lower birth weight infants, effects on the infant's mental development, and changes in gene expression that potentially affect risk of disease later in life.[63] Lead exposure and lead poisoning, therefore, while often causing acute health problems in children, can contribute to serious life-long health effects and even intergenerational damage.

Dr. Reeds reported his findings to the Panhandle Health District, the local health department, which sent inspectors to visit the family's home in February of 1974 in hopes of finding the source of the children's lead exposure. They found nothing obvious and apparently did not consider the smelter to be a potential source of the children's exposure because the family lived about a mile from the stack.[64]

Reeds continued to monitor the Taylor children's blood lead levels. In early April 1974 another test showed that both children's blood lead levels had again increased. One child's blood lead level was nearly 70 µg/dL and the other's was close to 90 µg/dL. The children were hospitalized and underwent chelation treatment. After five days of treatment their blood lead levels were closer to 40 µg/dL, and the children were discharged.[65]

Around the same time, Dr. Reeds tested for lead exposure a new mother and her infant who lived in the same area as the Taylor family. Reeds recalled that both the mother's and the baby's blood lead levels were "inordinately elevated."[66]

While the Taylor children were hospitalized, Dr. Reeds pushed the health department to do a more thorough investigation of Taylor family's home environment in the hopes of finding the source of their lead exposure. Investigators tested soil from the yard, dust from the front porch, paint from the bedrooms, water samples, and plumbing (lead can leach from older pipes), and they examined the children's toys. The highest lead levels were found in dust from the front porch, which was over 2 percent lead.[67]

It would later become clear to public health investigators that the steady rise in the Taylor children's blood lead levels between November of 1973 and April of 1974 was very likely due to the exceptionally high emissions from the Bunker Hill smelter during this time period.

George Dekan and Ian von Lindern had been warning of a possible public health and environmental crisis since at least January of 1974, due to the dramatic increases in lead in the air that they had detected. Five months after the first of their warnings, in late May of 1974, Dr. James A. Bax, Idaho's DECS administrator, wrote a sharply worded letter to Bunker Hill calling the company's attention to the possible health consequences of its emissions, noting the two cases of lead poisoning (the Taylor children) that had already been reported, and asking the company to provide data on emissions and production levels.[68]

Just a month earlier Bunker Hill had written to land owners in the Kellogg area to warn them about the hazards of pasturing livestock, particularly horses, in the vicinity of the smelter due to the potential for lead poisoning. Although the letter blamed the build-up of smelter emissions in soil over sixty years, recent increased emissions may have been a factor in the timing of the warning. Landowners were advised not to pasture their animals in certain areas due to high levels of lead in soil and pasture grasses. The letter reassured residents that the build-up of lead in soil was "not dangerous to humans."[69]

Bob Cutchins was a miner who began working in the Silver Valley in the early 1950s. In the late 1960s he married Sarah and worked in the infamous Sunshine mine, missing by a few years the disastrous underground fire that killed ninety-one miners in May of 1972. For a while, tired of mining, Cutchins followed blue collar jobs around the western United States. But in the spring of 1972 he and Sarah wound up back in the mining district; Cutchins would once again be working hundreds of feet below the earth's surface.[70]

By this time Bob and Sarah had two young children—Emily and Robert. The family rented an old, wood-framed house west of Kellogg and about one-quarter mile from the Bunker Hill smelter's stack. A picture of the house taken in 1974 shows very little vegetation in the yard. Soil samples from the area taken by Bunker Hill in 1975 revealed very high lead concentrations—between 8,000 and 11,000 parts per million—much higher than current environmental standards permit. The Cutchins family would live in the house near the smelter stack for about a year and a half, until shortly after the family was told to immediately move the children away from the smelter's towering smokestacks.[71]

Bob and Sarah's third child, Stacey, was born while they lived near the smelter stack. She was often cranky and suffered from fevers, headaches, and cold symptoms. Her legs seemed to bother her, and she appeared to have trouble with her balance. All three children had similar symptoms. The oldest, Emily, would complain of leg pain and would frequently squeeze her head, indicating that it hurt. Robert, who was a baby when the family moved back to the Silver Valley, began suffering from headaches and stomach cramps during the time they lived there. Too young to talk, he would pull at his hair to indicate his distress. The children also had recurring fevers and flu-like symptoms.[72]

According to Cutchins, despite multiple doctor's visits for evaluation and treatment of the children's health complaints, the first he learned that his children were suffering from lead poisoning was late in the summer of 1974 when the family was told by CDC employees "to get the kids to a hospital immediately." All three children had blood lead levels over 100 μg/dL.[73]

Cutchins learned that his children had been poisoned because local, state, and federal public health agencies mounted a study of lead exposure among children living near the Bunker Hill smelter in the spring and summer of 1974. After the Taylor children were hospitalized and very high lead levels were consistently measured in ambient air and in environmental samples taken near Kellogg, the parallels with El Paso were obvious. Based on the levels of contamination in the environment, officials suspected that the problem near Bunker Hill would be worse, and they were right.

The study came about because a commissioned officer of the U.S. Public Health Service, Dr. David McDaniel, was on a two-year assignment to the Idaho State Health Department. McDaniel learned about the Taylor children and told Philip Landrigan, who was still employed by the CDC. Landrigan and the CDC contacted the leadership of the Idaho State Health Department and began to discuss the need for an investigation.[74] The Panhandle Health District and State Health Department, Philip Landrigan, and others from the CDC began planning the blood sampling and environmental study in May of 1974.

By mid-August, federal, state, and local health officials were going door-to-door looking for children, drawing blood, and testing dirt and dust in increasingly distant areas from the smelter. Area I, within one mile of the smelter, included Smelterville and Deadwood Gulch. Area II, one to two and one-half miles from the smelter, included Kellogg. Other towns such as Pinehurst, Big Creek, Osburn, Wallace, and Mullan were included in the outlying areas.[75]

By early September of 1974 the preliminary findings were available. Robert H. Allen, Gulf's CEO, learned of their imminent release from a confidential telegram from Bunker Hill's Kellogg legal team, which was monitoring the situation. It read in part: "We expect that the state of Idaho will release this afternoon a statement on lead intoxication in Kellogg, Idaho with adverse implications for Bunker Hill's operation."[76]

The findings of the blood tests were alarming. Of the 919 one- to nine-year-old children tested, 42 percent had blood lead levels between 40 and 79 µg/dL, and 4.5 percent (41 children) had blood lead levels over 80 µg/dL. Within a mile of the smelter, 98.8 percent of the children had blood lead levels over 40 µg/dL, and 38 children had lead levels over 80 µg/dL. Levels were highest in one- to four-year-olds—and the mean in this age group in the area closest to the smelter was 72.5 µg/dL—a level that can result in serious long-term health consequences.[77] The highest blood lead level of 164 µg/dL was measured in a fourteen-month-old baby. Dr. Bertram Carnow, who helped to uncover the El Paso lead poisoning problem, would call three of the children with blood lead levels over 100 µg/dL "walking dead babies" since children who experienced blood lead levels that high were at risk of death.[78]

Health officials visited the homes of the children with blood lead levels over 80 µg/dL and told parents to take their children immediately for medical treatment. Medical treatment for many in Kellogg meant taking their children to the Doctor's Clinic, which had a financial relationship with the Bunker Hill Company. Two of the clinic's physicians publicly stated their beliefs that fears of lead poisoning were overblown.[79] And if children were hospitalized, they would probably go to West Shoshone Hospital, where Doctor's Clinic doctors would treat them.[80]

As the lead poisoning epidemic became public, Dr. Panke, quoted in the press criticized state health officials, saying, "I feel some panic was created.

There wasn't any emergency situation." Panke went on to state that lead intoxication was treatable.[81] Dr. Robert Cordwell, a Doctor's Clinic physician and president of the Shoshone County Medical Society, said that the health department had prematurely released the blood test results, which were "blown out of proportion, causing a crisis panic."[82] In a *People* magazine article Cordwell was quoted as saying "treatment is very simple" and that "it's up to the mothers—they've got to teach children not to eat dirt and to wash their hands."[83]

The claim that parents, or mothers more specifically, were at fault if children were "leaded" was one that the lead industry had made for years—that lead poisoning was attributable to poor parenting. This would become a key part of Bunker Hill's defense when families subsequently sued the company. Witnesses and the company's attorneys would argue in court that the reason for some of the children's problems was that their homes were filthy and that their parents used drugs. In fact, claims such as these may still be influencing parental actions in the Silver Valley. A recent study found that such stigma remains a barrier to children participating in ongoing blood lead sampling in the area.[84]

The idea that lead poisoning was easily treatable was not supported by evidence either then or now. The standard treatment for lead poisoning was then and still is chelation, which involves the administration of chemicals that bind with lead and other heavy metals in the bloodstream and allow them to be excreted through urine. Chelation can reduce blood lead levels, but it is not clear if such treatment can prevent the tragic long-term effects of lead poisoning such as cognitive and neurobehavioral problems. And chelation therapy is not without risk; children must be carefully monitored, and the treatment can be painful and produce side effects. Regardless of whether children are treated or not, it is thought that lead's effects on children's nervous systems are irreversible.[85]

In the early days after the crisis was revealed, Bunker Hill/Gulf said little to the press although it did issue a statement expressing surprise and asserting that, if the test results were found to be correct, it had "complete confidence in community doctors to effectively treat any cases of lead poisoning actually encountered."[86]

It appears that Bunker Hill/Gulf may have used the data collected by the CDC to estimate its potential financial liability for poisoning children in the Silver Valley. An undated handwritten memo sketched out the possible costs it might incur based on ASARCO's costs in El Paso. The memo in parts reads, "El Paso—200 children, $5 to 10,000/kid + land purchases + associated claims. Bunker Hill—500 kids over 40 µg + 40 over 80 µg—6–7 million."[87]

Behind the scenes Bunker Hill/Gulf was strategizing with power brokers in the smelting and lead industries to craft a response that would allow the company to achieve several aims: weather the crisis, keep the smelter operating, exercise control over the research that was conducted in the community and on the affected children, and limit the cost of treatment.[88]

Meanwhile, the Lead Industries Association (LIA) and its public relations firm Hill & Knowlton were tracking the Bunker Hill crisis because of possible negative consequences for the lead industry as a whole. Reinforcing the centrality of industry-sponsored research to respond effectively to these types of crises, a Hill & Knowlton staff member suggested to the LIA as the news was breaking that since "another El Paso situation [was] in the making in Idaho, . . . a McNeil-type study in Idaho" might be initiated. The Hill & Knowlton staffer also suggested making a public statement on the McNeil work in El Paso in connection with the crisis in Idaho as "this might provide an opportunity for a news release that could help set this in perspective as far as the health implications go." The public relations firm was also looking ahead to the upcoming American Public Health Association annual meeting and was concerned that Philip Landrigan might use this forum to present the CDC findings on Silver Valley blood lead levels to the public health community. The Hill & Knowlton staffer wrote, "This would make it doubly important that someone representing LIA be in attendance and use whatever opportunity presents itself to bring up the McNeil work."[89]

After the crisis became public, Bunker Hill and ILZRO leadership initiated discussions regarding the help that ILZRO could provide. Jerome Cole planned a visit to Kellogg to meet with Gene Baker in mid-September 1974. The only problem was that Bunker Hill was not then a member of ILZRO.[90] By October Bunker Hill/Gulf had officially joined, and ILZRO's executive vice president promised that his organization would "certainly do everything within our power to help you with this study in Kellogg."[91]

Bunker Hill/Gulf also met with ASARCO managers and lawyers to "glean experience from the El Paso lead problem which could be translated to the Kellogg situation." Based on that meeting Bunker Hill determined it would keep the smelter operating (as ASARCO did in El Paso, it would argue that current emissions were not the problem), try to reduce lead emissions, and buy and destroy the homes that were close to the smelter and relocate residents. ASARCO also reportedly advised Bunker Hill to "stay clear of Dr. [Philip] Landrigan." According to a top Bunker Hill manager, Landrigan was "paranoic about proving the existence of adverse health effects at very low blood lead levels and was not honest in his efforts to prove same." Bunker planned to mount what it called an "independent study through a collaboration with ILZRO."[92]

Bunker Hill was opposed to allowing Landrigan and the CDC to lead an unfettered investigation of the problem in Kellogg. Years later, Robert Allen, Gulf's CEO, said in a deposition in reference to letting the CDC supervise, "obviously we objected to that from the very beginning." Allen said he thought the CDC was "biased."[93] In the coming years the State Health Department did not allow the CDC to return to Idaho to lead or participate in public health investigations during critical years.[94]

Without the outside oversight of the CDC, there was no independent party to ensure that children were being adequately monitored and treated, and the state and Bunker Hill publicly denied that children had been permanently harmed.[95] This prevented parents and the community from fully understanding the magnitude of the problem and very likely prevented children from getting treatment and help with inevitable developmental and health consequences. Entirely preventable exposure and poisoning continued to occur through the early 1980s in children of long-time residents and among families who moved into the area. For the children of the Silver Valley, the response was nothing short of tragic.

Early in its response to the crisis, one of Bunker Hill's main aims in collaborating with ILZRO was to establish an intermediary who could refer children for medical attention—and pay for it—so that the company would not be directly involved in the children's care. In part, Bunker Hill was concerned that direct payment for medical care might amount to an admission of responsibility "in a later negligence or nuisance action." In an early October 1974 meeting, Gulf and Bunker Hill lawyers hashed out the details with ILZRO officers Schrade F. Radtke, Jerome F. Cole, and Donald R. Lyman and ILZRO counsel. ILZRO would conduct a study that would combine research "to ascertain the effect of an elevated blood lead level" with medical referral, and would be the company's intermediary, paying for medical treatment (Bunker Hill would reimburse ILZRO). Bunker had two stipulations to impose on anyone who received care: "No treatment would be provided unless the person participated in the study, and there would have to be some ceiling or limit on the treatment."[96]

Liability was not Bunker's only concern with respect to paying for medical care for lead poisoned children. Since many of the affected children would likely be covered under Bunker's medical plan with the Doctor's Clinic, providing hospitalization and other treatment for them out of the existing plan might cut into the doctor's salaries. Notes from the meeting summarized Bunker Hill attorney William Boyd's statements. He explained to the ILZRO representatives at the meeting: "if the cost of treatment for children came out of this program, the doctors would be upset." And the company was "anxious not to disturb the doctors."[97]

The plan to have ILZRO conduct a follow-up study in the Kellogg area hit some snags. For one, federal and state agencies were also planning a comprehensive investigation around the Bunker Hill smelter. Dr. Bax of the State Health Department was upset by the prospect of ILZRO conducting a study, and he made his concerns public. Bax told the press that he would "hate to see these

Kellogg area children being used as guinea pigs in a mining-industry sponsored study" and alleged that the ILZRO-backed El Paso study (the McNeil study) was terribly flawed.[98]

State officials were well aware of the controversy that the McNeil study had caused in El Paso and wanted to avoid this in Idaho. They urged community members not to participate in any studies conducted by ILZRO. A local health department official saw Bunker's use of ILZRO as "purely a defensive move ... to defend themselves from any future lawsuits ... because most of the findings of this organization are often different than those of other researchers." But some Kellogg doctors were described as feeling differently, supporting the involvement of ILZRO and questioning the "objectivity of the state and federal agencies."[99] Dr. Panke had already agreed to work with ILZRO on the planned study and told Jerome Cole that he was "convinced that you and your organization are dedicated to honest and open research."[100]

Bax decided to try to come to an agreement with Bunker Hill about how to proceed. He proposed that the company and the state collaborate on a single study in exchange for Bunker's abandoning the ILZRO study. His proposal required that Bunker Hill improve environmental conditions and provide study funding. What Bax might not have realized at the time was that this arrangement worked much more to Bunker Hill's advantage than to the state's. Bunker Hill had a hand in selecting the study's leadership and shaping the scope and approach of the study while at the same time enjoying the credibility of collaborating with the state, rather than ILZRO.[101] The joint state–Bunker Hill study was called the Shoshone Project. ILZRO, however, remained involved behind the scenes and continued to provide assistance to Bunker Hill.[102] Communication with ILZRO came both from Bunker's management and possibly from Dr. Panke, who told Jerome Cole early in the crisis that he would "attempt to keep you informed of the situation as it develops here, and of course, I would do this on a purely informal basis, as from one friend to another."[103]

The Shoshone Project was to have a technical committee, a group of experts who would make decisions about research and medical treatment. A planned CDC follow-up study would continue but would be managed by the technical committee of the Shoshone Project. The CDC had little say in this decision because it conducts investigations in states and counties at the invitation of the respective officials. Bunker Hill pushed for this arrangement because it felt that this "would force [Philip] Landrigan to be honest."[104]

To get recommendations for the Shoshone Project's technical committee, Bunker Hill/Gulf initially turned to Dr. Raymond Suskind, the director of the Kettering Laboratory at the University of Cincinnati, which had long been funded by the lead industry. Suskind suggested a number of lead industry–affiliated researchers, including Sydney Lerner, a Kettering researcher and medical consultant for the industry; Paul Hammond; and George Roush, Monsanto's medical director who formerly held the same position for the Ethyl Corporation.[105]

Ultimately Bunker provided a different list of experts to the state for consideration for Shoshone's technical committee. These experts had either received lead or smelting industry funding or were associated with industry interests including J. Julian Chisolm, Donald Barltrop, Henrietta Sachs, James L. McNeil, Philip Enterline (a University of Pittsburgh scientist who had conducted occupational health studies for ASARCO and evaluated studies for ILZRO), Jerome Cole of ILZRO, and Gary Ter Haar of the Ethyl Corporation.[106]

In the end the technical committee was composed mainly of physicians and professors from the University of Washington's (UW) Medical School and School of Public Health, with a few outside members. A few members are known to have had relationships with the lead or smelting industries, including J. Julian Chisolm, Ronald Panke, and Ted Loomis, a UW professor of pharmacology and Washington State toxicologist. Loomis had testified on ASARCO's behalf at a public hearing in Tacoma on health hazards to children from the Tacoma smelter's arsenic emissions.[107]

Glen E. Wegner, a medical doctor and lawyer, was appointed jointly by the governor of Idaho and the president of Bunker Hill to be the Shoshone Project's director. Wegner was a well-connected Idahoan and a former White House fellow, special assistant to the surgeon general, and deputy assistant secretary for health legislation in the Nixon administration. According to a Bunker Hill executive, after completing the Shoshone Project, Wegner would go on to consult for the Doctor's Clinic on Bunker Hill business.[108]

The Shoshone Project was said to be "independent of both The Bunker Hill Company and the State of Idaho" but Bunker Hill would pay for the study (initially estimated at $250,000–$300,000) and Wegner's $5,000 monthly salary. At times Bunker Hill managers seemed to treat Wegner as if he were their employee and requested detailed information about study progress and decisions. For example, in November of 1974 Gene Baker asked Wegner to provide him with information on all tasks that the state was planning and its collaboration with federal agencies, all data that had been collected, information on all blood lead levels by age and address, and copies of notes taken by the health department on family living conditions and home environments.[109]

Wegner seemed to command respect in Idaho, and few people publicly criticized his leadership. Staff in the State Health Department, however, internally questioned his handling of the project and his relationship with Bunker Hill.[110] Bunker Hill/Gulf's public relations firm warned internally: "Dr. Wegner is an asset, but one that could be destroyed if his independence becomes suspect."[111]

The memorandum of understanding between Bunker Hill and the state for the Shoshone Project limited its scope to the "lead health problem" in the community.¹¹² No other toxic metals would be considered, even though cadmium exposure was also a concern of health officials. Arsenic and mercury were also present in significant quantities the smelter's emissions.

The stated aims of the Shoshone Project were to provide medical attention to children with blood lead levels over 80 μg/dL; evaluate children with blood lead levels between 40 and 80 μg/dL and "with the combined use of personal hygiene and environmental manipulation of home and surrounding environment, reduce these levels below 40"; develop a lead protocol that will lead to a long-term acceptable solution to the environmental lead health problem in the valley"; and "determine if there were any long-term health problems generated by the existence of elevated blood lead levels."¹¹³

There were several important aspects of the Shoshone Project that were not in keeping with the surgeon general's 1971 childhood lead exposure treatment guidelines. Specifically, the decision to provide treatment only to children with blood lead levels over 80 μg/dL was inconsistent with the standard of care at the time. Children with blood lead levels between 50 and 79 μg/dL should have been medically evaluated for lead poisoning, and those with clinical symptoms such as lead lines, central nervous system disorders, and constipation, among others, should have been treated immediately. This group of children also should have been closely monitored to make sure that their blood lead levels did not rise.¹¹⁴ The Shoshone Project's protocol did not reflect the urgent concern of many in the medical and public health communities for children with blood lead levels over 50 μg/dL in the early 1970s.

The surgeon general's recommendations also called for monitoring and testing children who had suffered lead poisoning until at least age six "to prevent repeated exposure or poisoning," and for ongoing neurological and behavioral testing for several years because damage from lead poisoning may not be immediately apparent.¹¹⁵ Silver Valley children with the highest blood lead levels—those over 80 μg/dL—were not followed in any systematic way by the state or anyone else in the ensuing years. Many of them were relocated by Bunker Hill and their homes were destroyed, and a number of the children moved out of the state. The State Health Department determined that "no attempt would be made to follow them" since "no permanent neurological damage could be demonstrated."¹¹⁶ The youngest children, who had the highest blood lead levels (those one to three years old in the early 1970s) and were most at risk, were not given follow-up tests as part of the Shoshone Project to determine if they had experienced developmental damage.¹¹⁷

Other omissions in the Shoshone Project's response to the lead poisoning crisis are also notable. There were apparently no advisories to pregnant women that their fetuses were at risk, and it does not appear that there were concerted efforts to prevent exposure and poisoning in this vulnerable group.

Infants in the Silver Valley were being exposed to lead at levels that could have lasting significance for their health even before they were born. One small study of mother–newborn pairs in the valley found average blood lead levels in newborns of 27.6 μg/dL between 1976 and 1978, two years after the period of maximum exposure.[118] By today's standards these infants would be considered dangerously lead exposed.

Bunker Hill's role in funding the Shoshone Project and its collaboration with the state gave it access, through Dr. Panke, the company's medical consultant to the project, to the medical test results of the participants. This of course, raised important questions of conflict of interest and confidentiality. The "study participants" were also potential plaintiffs in legal actions against Bunker Hill. Access to their medical records might unfairly advantage the company in future legal proceedings. Children whose parents brought a lawsuit against Bunker Hill were referred by the Shoshone Project for intensive medical evaluation.[119]

In making the referral to the UW Medical School, a local physician wrote: "All [number of children omitted] were heavily leaded in the Kellogg survey last year. 3 months after being away from the area, their blood leads & FEPs are not decreasing. Another course of EDTA +-BAL [chelation therapy] is probably in order. The [X Family] are trying to connect [one of the child's] [health issues] with lead, and have a lawsuit going against Bunker Hill.... The lead study group would like all aspects of the [X Family] children checked out thoroughly while they are in Seattle."[120]

Despite claims to the contrary, Bunker Hill monitored the activities of the Shoshone Project and tried to influence the project's internal decision making. For example, when a child with obvious physical and mental disabilities was included in an aspect of the study, Gene Baker wrote an exasperated letter to Dr. Wegner protesting the child's inclusion. The child's data was not included in the study. Children with the most severe health issues (which may or may not have been due to lead) may have been screened out or otherwise not included in the health evaluations.[121]

Three months into the Shoshone Project Wegner released an interim report to update the community on its progress and findings. He reported that one hundred children had undergone neurological and psychological testing and "no known serious effects on the children" were found. Forty-four children with blood lead levels greater than 80 μg/dL were being treated by a physician, twenty-two families had been relocated, and any family living within a half-mile of the smelter was eligible to sell its home to Bunker Hill. By January of 1975 forty-two of seventy-five homes had been bought and demolished. Bunker Hill was also making soil, sand, and gravel available to area residents that was said to be free of lead. The Silver King School Board had voted to close the school at the end of the year—a vote that was later reversed by voters with the support of the Parent Teacher Association. Those children who had undue lead absorption were being visited by home health workers and advised on how to keep dust levels in the home down.[122]

Wegner, as the official voice of the Shoshone Project, appeared at times in his public communications to venture into the territory of public relations for Bunker Hill, such as when he said that the company was committing to reducing "as rapidly as possible lead emissions from the smelter to a point at which there will not be even the remotest possibility that they can be affecting the health of valley residents."[123]

As the disaster unfolded, Bunker Hill made much of its intention to reduce lead emissions from the plant and claimed that it was making great progress in this effort.[124] This, however, was debatable. About six months after Wegner reassured local residents that Bunker Hill was working hard to reduce its lead emissions, von Lindern inspected the plant and noted visible fugitive emissions as well as visible emissions from the main baghouse, which "seemed to be operating poorly." Von Lindern thought that Bunker Hill was still pushing its equipment to boost production. He concluded succinctly: "we are sacrificing clean air for the sake of Bunker Hill production."[125]

As time went on, von Lindern became even more outspoken about his concerns about Bunker Hill's pollution and the state's response within the health department. In June of 1975 he wrote a memo to Lee Stokes, the head of the environmental services department within the State Health Department, bluntly accusing the department of "[speaking] from the same mouthpiece" as Bunker Hill. He charged, "Bunker Hill at times seems to be interpreting our regulations for us," and said the agency's credibility [with the public] was "below zero." Von Lindern summarized his concerns: "We have just had probably the largest industrially-sourced, heavy metal intoxication epidemic in the United States. Despite this definite health hazard, it appears to the public that we defend the company and deny the problem."[126] Von Lindern also contacted Philip Landrigan to alert him that the likely reason for the very high blood lead results in the August 1974 study was that Bunker Hill had bypassed the baghouse and substantially increased its lead emissions in the months before the CDC's testing began.[127]

In August of 1975 von Lindern even took his concerns to attorneys who were representing a family that was planning a lawsuit against Bunker Hill. At that time he told lawyers about flaws in the Shoshone Project, including his view that medical tests were being performed only on older children. Von Lindern also discussed with attorneys his belief that confidential health department records regarding the lead poisoning epidemic in the Silver Valley were winding up "in the hands of Bunker Hill" and ILZRO.[128]

Early in the crisis Bunker Hill hired two public relations firms to help it navigate the crisis and limit the potential damage. Bunker Hill aimed to improve its public image and gain and maintain the support of the local community and elected officials. At the same time, it was actively fighting state and federal regulation of their sulfur dioxide and lead emissions as well as more protective worker safety regulations.[129]

The public relations firms advised Bunker to be outspoken about its attempts to address the plant's pollution problems—both historic and ongoing. For example, they thought that Bunker should be publicizing the fact that it was giving away free soil and gravel to homeowners, "even though it wouldn't be needed if it weren't to be used to cover lead on the ground." Bunker embarked on a "greening program" in an attempt to plant trees and ground cover around the plant and surrounding hillsides where little grew. Bunker's revegetation efforts met with little success, likely because of accumulated heavy metals in soils toxic to plants and ongoing emissions from the smelter. Again, the public relations firm thought this should be publicized more and wrote, "even though the hillsides are still barren, there is much you can point to with pride."[130]

An opinion poll conducted for the company in early 1975 showed, however, that the majority of local residents remained firmly behind the company. Eight of ten Kellogg residents thought the company was doing "all it reasonably can against pollution;" and only one in eight had strong negative feelings about the company. Only 6 percent of Kellogg residents thought that pollution was a serious danger to health, compared with 25 percent of people in the rest of Idaho. The results helped to inform a campaign to "establish Bunker Hill as a leader in the fight against pollution," focusing on the local and state residents under age thirty-five—the group most likely to be concerned about pollution problems.[131]

Many community leaders were also solidly behind mining and smelting and supported Bunker Hill. Wendell Brainerd, the editor of the *Kellogg Evening News*, told representatives of Bunker's public relations firm that the "bad publicity" Kellogg was receiving was largely a problem of outsiders not understanding the mining and smelting industry. The company had strong support from other mining companies in the area, which relied on the smelter to process its ore. The vice president of Hagadone newspapers, which ran the *Coeur d'Alene Press*, reportedly told public relations representatives that the *CDA Press* "has been and will continue to be friendly."[132]

Claims that the lead problem in the Silver Valley was improving were often uncritically reported in the press. In the spring of 1975 the national press was reporting that lead levels in children near Bunker Hill had shown a "34% decrease."[133] And in the summer of 1975 Bunker Hill officials claimed that air lead levels had been "substantially reduced" between September 1974 and May 1975 with average ambient lead levels in the area of 5.5 $\mu g/m^3$.[134]

Both claims were questionable. Both ambient lead levels and blood lead levels were subject to seasonal fluctuation. Both were likely to be highest in drier months, such as August and September, which is when the 1974 blood lead testing occurred. Comparing these levels to levels taken in the spring of 1975 could make it appear that blood lead levels had fallen even though this could be attributable to seasonal fluctuation. The same was true for lead levels in the air.

Dr. Wegner, speaking publicly in the community during the summer of 1975, said that he was "cautiously optimistic" that Kellogg children had "not been substantially harmed by lead from the smelter." Wegner told the Idaho Mining Association, "We know there are no obvious, gross deficiencies in the children. We're looking carefully now at the subclinical data. But there is some evidence to say that if there was a problem, we've caught it in time and turned it around."[135]

Not everyone shared Wegner's optimism. An editorial in the *Idaho State Journal* was critical of his and the state's handling of the problem, charging that the state had not made the study's financial backing public, nor had it made it clear who was on the project's payroll and how much they were being paid. Of Wegner the editorial said, "Someone receiving $5000 dollars a month from the company to head the study isn't likely to say publicly the company is continuing to poison the atmosphere in the Silver Valley and endanger the lives of the children living there."[136]

The editorial also questioned the intent of state government to take action to protect health in the Silver Valley: "If it took [the State Health Department] a half century to figure out there is a lead problem in Kellogg, how long will it take the department to act if it ascertains Bunker Hill is fudging on the seriousness of the contamination? What will happen if it becomes a question of the state taking action to close the smelter? . . . The question is, who is watching the state watch Bunker Hill watch lead pollution in Kellogg?"[137]

This was a pertinent question in Kellogg because, unlike in Tacoma, where there were citizen environmental groups and organized outspoken opposition to the smelter from various quarters, there was much less organized opposition in Kellogg. In fact, the community voted overwhelmingly to keep the Silver King Elementary School open after the lead poisoning crisis was discovered, even after the school board ordered its closure. In urging parents to vote to keep the school open, the PTA pointed to the lack of "conclusive evidence that our children are being leaded by attending Silver King School."[138] There were parents who were very concerned about their children's health, but some found it difficult to speak out in a community where mining and smelting was so dominant. Without the countervailing pressure of significant organized community opposition, it was easier for Bunker Hill's views and those of the Shoshone Project to dominate the discourse about community health effects. It also led to less community oversight of the state's actions and allowed the state to align with industry's interests over the health of the public.

Because the CDC had demonstrated in its 1974 study that four hundred of the Silver Valley children tested had lead levels over 40 µg/dL, and 10 percent of those children had blood lead levels over 80 µg/dL, and based on what was known about lead's ability to damage children's brains and bodies, it would be hard to make a convincing case that some children in the Silver Valley would not suffer harm. Bunker Hill, however, tried to make this case on multiple occasions.

Reminiscent of McNeil's hypothesis regarding the El Paso children, early on in the crisis a company official made the claim that lead poisoning in Silver Valley children was somehow physiologically different than lead poisoning in urban children. At a Mining Association Convention in December 1974 Gene Baker stressed the uniqueness of the exposure in the Silver Valley. Even though many Silver Valley children had lead levels high enough to suggest they should have clinical symptoms and permanent damage, both Bunker Hill and the State Health Department would maintain for years that they did not.[139] Baker suggested that the reason "for the apparent lack of symptoms in even the cases of highest reported lead absorption" might be that Silver Valley children were exposed more gradually by breathing lead in air than a child who poisoned by sudden ingestion of lead paint.[140]

In the late 1970s and early 1980s, the State Health Department would further this argument, positing that other air toxics emitted from the smelter might actually have a protective effect on children, preventing them from becoming damaged by lead. It even proposed a study be conducted to identify "substances which may be protecting against clinical lead poisoning."[141]

With Bunker Hill and the State Health Department exercising a large measure of control over the research that was done, the community did not have the opportunity to learn about community health risks from outside researchers. One of the few independent studies was a small study conducted in the fall of 1974 by Philip Landrigan and the CDC aimed at assessing damage in exposed children. In a case control study comparing five- to nine-year-old children with blood lead levels over and under (controls) 40 µg/dL, a statistically significant difference was found with respect to nerve conduction velocity (NCV). Children with higher blood lead levels had slower nerve conduction velocity (which suggests damage to the nervous system), and NCV was shown to decrease with increasing blood lead levels.[142]

This paper was published as part of the Shoshone Project's findings along with a criticism authored by Dr. Panke and P. S. Gartside, an assistant professor of biostatistics at the University of Cincinnati Medical Center. The critique called into question the study design and statistical techniques, and pointed to possible sources of error in the study. Gartside and Panke also reported that six of the children with the lowest nerve conduction velocities were retested at the UW Medical School and finding normal NCVs, "thereby indicating reversible nerve conduction velocity depression, or possibly technique differences or recording error in data collection."[143]

When the Shoshone Project's findings were released in early 1976, Landrigan's NCV findings were downplayed; children with the highest exposure to lead were said to have "a higher than expected prevalence of mild anemia and had minimal degrees of slowing in nerve conduction." The main conclusion of the Shoshone Project's committee was "we do not feel any permanent clinical impairment or illness has occurred. Further, it is not likely to occur in the future due to this particular exposure."[144]

But there was apparently a dissenter on the committee, Dr. Ralph Reitan, a professor of psychology and neurological surgery at the University of Washington. He was the developer of Reitan testing, a neuropsychological examination that was used to assess a small number of lead exposed Silver Valley children. Eleven children had been referred to Reitan for testing by the Shoshone Project.[145] In a personal communication with Philip Landrigan, Reitan apparently stated that the preliminary tests of children with the highest blood lead levels showed "subtle abnormalities in central function, even in the face of normal intellectual capacity."[146] Reitan's statement however, was not included in the Shoshone Project's final report.[147]

When the Shoshone Project was about to release its findings, one of Bunker Hill's public relations consultants suggested that Glen Wegner and Gene Baker meet with him to "prepare the Bunker Hill statement and press release." Further, the public relations consultant wrote, "whatever conclusions the report may reach, it would seem to be important that its publications establish ... in print evidence that there is no basis for suit against Bunker Hill for reason of damage to health."[148]

Some State Health Department staff thought at the time that a concerted effort was being made to keep Philip Landrigan from having significant influence on the Shoshone Project's findings. One staff member alleged that Landrigan was prevented from attending some meetings of the project's technical advisory group and that Landrigan had been told by Glen Wegner not to come back to Idaho. In a 1978 interview with lawyers for lead poisoned children, she said that health department staff "joked about Wegner standing at the airport with a loaded shotgun."[149] When the case got to trial, lawyers for lead poisoned children would allege that an Idaho state senator pressured the State Health Department to prevent Landrigan from testifying at the 1981 trial.[150] In the end, Landrigan did testify for the plaintiffs.

As questions arose in subsequent years about the impact of the disaster on Silver Valley children, both State Health Department officials and Bunker Hill leadership would point to the findings of the Shoshone Project as evidence that children had not been harmed.[151]

Philip Landrigan, for one, took issue with the Shoshone Project's findings and spoke publicly about his concerns in subsequent years. On several occasions in the press he charged that its purpose was to "find there was no problem." He also accused the state of Idaho of "frank conflict of interest" for its participation.[152] In 1980 he said of the Shoshone Project, "It became clear to us that the purpose of this whole [project] was to blunt the federal government's findings" of health damage in exposed children.[153]

After the first round of blood lead testing in 1974, Silver Valley children not surprisingly continued to have elevated blood lead levels. Ambient concentrations of lead were still high, and soil and house dust near the smelter was highly contaminated. After 1974 the CDC was briefly involved in blood lead testing in 1975, but beginning in 1976 the State Health Department and Bunker Hill took control of annual testing.[154] Their approach relied on a convenience sample of children rather than a random sample that would have yielded a more valid estimate of the problem. Parents were also aware that Bunker Hill was involved in the testing and had access to the results, which may have been concerning for some, particularly employees of Bunker Hill. Also, because the Shoshone Project reported that children had largely not been harmed, parents may have reasoned that if lead exposure in the Silver Valley was not resulting in permanent harm, what was the point of having one's child tested?

But the blood lead results among those tested continued to alarm CDC officials. Closest to the smelter, a high percentage of children had blood lead levels over 40 μg/dL throughout the 1970s, though smaller numbers of children were tested than in the original 1974 study. For example, 70 percent of those tested in 1975 had blood lead levels over 40 μg/dL, and 52 percent had levels this high in 1978. Farther from the smelter, 44 percent of children had blood lead levels over 40 μg/dL in 1975, 32 percent in 1977, and 21 percent in 1980. Some children who moved to Smelterville after 1974 were found to have blood lead levels up to 70 μg/dL.[155]

Elsewhere in the United States average blood lead levels were declining as lead was being phased out of gasoline. Between 1976 and 1980 the median blood lead level in U.S. children between ages one and five was 15 μg/dL and about 4 percent of children between six months and five years had blood lead levels over 30 μg/dL.[156]

Participation in blood lead testing waned over the years. By 1979 only twenty-eight children took part in the annual August screening. Among this group, twenty-one had blood lead levels over 40 μg/dL, reflecting a possible increase in blood lead levels.[157] Bunker Hill attributed the increase to "many new parents in the area [who] do not fully understand the value of preventative health measures" as well as "reduced concern."[158]

Philip Landrigan and the CDC were unable to return to the Silver Valley to lead efforts to study the lead problem through the early 1980s. Landrigan testified in November 1979 to a Clean Air Act oversight committee that the CDC had been asking state officials to allow it to do a follow-up study of children's blood lead levels since April of 1978 and it had yet to receive a reply.[159] When it learned of the apparent increase in blood lead levels in 1979, the CDC pressed the state with greater urgency.

Finally, in March of 1980, Edward Gallagher, the state health officer, invited the CDC to conduct follow-up neurological testing on children who had been found to have blood lead levels over 80 µg/dL to determine if they showed evidence of brain damage. Since 1974 at least forty-seven children had had blood lead levels this high, and the State Health Department believed there were more who met this criterion living in the community in 1980 but had not yet been identified. Also, eighty children had been born in Smelterville (very close to the smelter) between October 1972 and March 1975, but the health department had "very little blood lead data on this group." This group of children lived closest to the smelter during the baghouse fire and aftermath and would likely have been the most highly exposed. But the State Health Department would not allow the CDC to assist with blood lead testing in the community; the CDC was told that Idaho had all necessary resources.[160] Instead the health department planned a blood sampling for April 1980 with the financial support of Bunker Hill.[161]

After the health department told Bunker Hill that the CDC had been invited to conduct neurological testing on children with blood lead levels over 80 µg/dL, Bunker Hill offered the state a $120,000 grant to "assess and eliminate potential lead/health problems in the Silver Valley and do additional medical testing of children if warranted."[162] Two days later the state informed the CDC that its help with neurological testing would not be needed since "we have developed the resources to accomplish all the neurological evaluations . . . [and] we feel it is no longer necessary for you to participate in the implementation of that portion of our scheduled activities."[163]

Dr. William Foege, the assistant surgeon general of the United States, sent an exasperated letter to Edward Gallagher, Idaho's state health officer, criticizing Bunker Hill's involvement and financial support. He warned that if the state went ahead without the CDC on-site, it would "be in the position of collaborating with a chronic polluting industry whose emissions you are supposed to be controlling." He also reiterated the seriousness of the situation in the area: "Although we have no wish to be alarmists, we must point out to you . . . the Silver Valley is the site of the worst community lead exposure problem in the United States. Neither in any large city where children ingest lead paint, nor in the vicinity of any other smelter, does the severity of lead exposure approach that found in Kellogg." Foege implored state officials to reconsider their decision not to allow the CDC's participation, but they did not.[164]

In 1981 Dr. Fritz Dixon of the State Department of Health explained the decision to not allow the CDC to participate in blood screening: "The whole issue became emotional. I had a lot of people say that if the CDC came, they would do all they could to see no one participated."[165]

Foege had also warned the state that conducting blood tests in April of 1980 would produce data that would not be comparable with previous years of sampling. In all previous years, testing occurred in August. August was the dustiest season in northern Idaho, when blood lead levels would be expected to be highest. Testing in April would be likely to show lower levels. Against the CDC's advice, the state conducted the testing in April of 1980.

Of almost four hundred children ages ten and under who were tested, forty-one children had blood lead levels over 40 μg/dL, and the highest value was 63 μg/dL. In reporting the findings, the state wrote: "Blood lead values in April of this year are definitely lower than the blood lead values previously obtained annually in August. It is not our intent to make a direct, all-inclusive comparison between blood lead values obtained in two different months." In the report on the April screening, the state emphasized that lowering blood lead levels was an "individual and community responsibility." Individuals were responsible for making sure children's hands were washed, and that they didn't put things in their mouths. The community effort did not involve Bunker Hill—rather it involved business owners providing clean and conveniently located restrooms so that families would be more likely to wash their children's hands frequently.[166]

On the basis of research conducted on subclinical lead poisoning in the early 1970s, the CDC had lowered the cut-off for "undue lead absorption" in children, and public health officials had become concerned about the effects of lead exposure at lower levels. By 1978 the definition of undue lead absorption had been lowered from 40 μg/dL to 30 μg/dL.[167] But the change in the definition was not incorporated into Bunker Hill's or the state's response to the community lead problem around Bunker Hill. In 1980 they were still using 40 μg/dL as the cut-off below which they aimed to get children's blood lead levels.[168]

Bunker Hill benefited from its collaboration with the State Health Department. In 1979, a year when parents seemed to be publicly challenging the findings of the Shoshone Project, the state health director, Edward Gallagher, was still repeating the contention that children in the area had not been harmed. At a community meeting to announce more blood lead testing in the area, Gallagher was quoted in the press as saying, "We don't know of any child in this area who has suffered from high lead levels." Further, he openly wondered why it was that children in northern Idaho "apparently do not suffer from lead in the air." He was also quoted as stating that children were being exposed to lead through ingestion—echoing the lead and smelting industry's position. Gallagher made this claim despite the fact that the average air lead level at the Silver King Elementary School in the last quarter of 1979 was over 10 μg/m^3—nearly ten times the lead limit of 1.5 μg/m^3 that EPA had set in September of 1978. Bunker Hill contended that plant upgrades completed in the mid-1970s (taller stacks) had reduced pollution problems, but the EPA contested this assessment and said that the pollution was now spread over a wider area.[169]

Even when confronted by outsiders about problems that children were reportedly having, Idaho health officials continued to deny that damage had been done. When interviewed for an NBC news segment, state epidemiologist Dr. John Mather was asked about learning difficulties that had been reported in lead poisoned children. Mather responded that "in any population there are people that are having learning difficulties, in any school district, in any place."[170] In a summary of a legal deposition from 1980 Mather apparently stated that "he believes that no children in Shoshone County have neuropsychological deficits attributable to lead."[171]

Meanwhile the State Health Department was planning the Bunker Hill–funded study of neurological effects on poisoned children. This study was slated to involve "detailed medical and developmental examination," but the testing would only be offered to children who were residents of eastern Washington or northern Idaho. Since many of the children with the highest blood lead levels had been relocated, some would be excluded because of this requirement. The results of the testing would be shared with the health department and the Bunker Hill medical consultant, Dr. Panke. Few of the children who had had blood lead levels over 80 μg/dL applied to have the evaluations done.[172]

Bunker Hill also directed the health department to "decline to offer this service" to two families who had children who would have been eligible on the basis of their blood lead levels in 1974, but they were involved in litigation against the company.[173] If the CDC had conducted the follow-up neurologic testing, children involved in litigation against the company would probably not have been excluded. Of eleven lead-exposed children tested by a professor of psychiatry at the University of Washington, seven "performed in the impaired range" on the Category Test used to assess some aspects of brain function, and as a group the children performed "very poorly."[174]

In the mid-1970s, while all of this was unfolding in northern Idaho, the lead industry, with coordination and funding from ILZRO, was engaged in a last-ditch effort to deny the health effects of airborne lead in the face of growing calls for its regulation. Whether children had been injured in El Paso and near Bunker Hill remained central to these discussions. For this reason ILZRO was intensely concerned with the outcome of the Muir Commission, which was evaluating the Landrigan and McNeil studies in El Paso. When one of ILZRO's grantees, Dr.

Clark Cooper, who was on the commission, was asked to write the summary and conclusion section of the commission's report, ILZRO deputy director Jerome Cole wrote to McNeil and said, "I think this is good and all we could have hoped for."[175] ILZRO was also kept apprised of the commission's discussions through Cooper.[176]

When California was reevaluating its air lead standard of 1.5 μg/m³ in 1975, ILZRO sent James McNeil and Dr. Panke to testify regarding the El Paso and Kellogg crises and to cast doubt on the CDC's research that children in these communities had been harmed by high concentrations of lead in air.[177] Commenting on all of the testimony from key players in the El Paso and Kellogg lead poisoning studies, ILZRO wrote, "The weight of the evidence strongly argues that no neurologic or intelligence deficits were produced by relatively high lead exposure in either El Paso or Kellogg. The findings in El Paso and Kellogg clearly demonstrate that there are no adverse health effects with blood lead concentrations considerably above 40 μg/100 g. California children in general have much lower blood leads which are in the normal range. Clearly this comparison indicates a significant margin of safety for California children."[178] ILZRO also highlighted the testimony of Henrietta Sachs, a researcher who consulted for the trade group and who discounted the threat of lead in air, stating, "If there were no lead in the air at all, the test results of inner-city children would not change."[179] The industry hoped to prevent the adoption of stringent federal standards for lead in ambient air and to roll back California's tough standard.

Although the EPA had determined in 1972 that air lead levels greater than 2 μg/m³ could cause "adverse physiologic effects" and "constitute endangerment of public health," the agency backed away from setting an ambient standard at the time after significant push back from industry.[180] Instead the EPA ordered a phase-out over time of lead in gasoline, hoping that the phase-out would result in ambient lead levels that were protective of health.[181] For years the EPA maintained that an ambient standard was not necessary because the phase-out of leaded gasoline addressed the predominant source of lead in air. The agency thought stationary sources (including smelters) could be regulated on an individual basis. Smelters, of course, were the most significant stationary sources. In the mid-1970s lawsuits forced the agency's hand. In March of 1976 a United States district judge in New York ordered the EPA to list lead as a pollutant.[182]

Bunker Hill was carefully tracking all of these developments because it was certain that the smelter could not meet an ambient lead standard of even 5 μg/m³. Baker reported to internal managers that he doubted the EPA would move quickly on setting an ambient standard even after being ordered to list lead. He noted, "EPA has said that promulgation of a lead standard is 'two lawsuits away,' one to force them to develop criteria and one to force them to issue standards."[183]

The EPA did appeal the court's decision to require the agency to list lead, but its appeal was denied.[184] In 1978, when the EPA announced its intention to set an ambient standard of 1.5 µg /m^3, Robert Allen, Gulf's president, slammed the EPA, charging the agency with the "pursuit of arrogant regulations," and he told Silver Valley residents that Bunker's fate rested not with Gulf but "with the federal government." "We are clearly trading capitalism for socialism in the country at an accelerated pace. We agree, I think, that it's tragic."[185]

In 1979 smelting companies were discussing among themselves the best strategies for "deferring expenditures" on the EPA lead standard, which meant extending the timeline for compliance. Bunker Hill, with the support of the state and apparently the EPA, was working on developing a plan to meet the standard "on paper" that was not "too onerous and has maximum expenditure delay." St. Joe's (a lead smelting company) wanted to get Congress to pass an appropriations rider that would "prohibit EPA for a period of 1 year from approving, disapproving or promulgating state plans for implementing the standard."[186]

State implementation plans for meeting the lead standard were due by July of 1979, but few states submitted them, Idaho included.[187] Region 10, however, was apparently moving ahead with an implementation plan for Bunker Hill. Gulf president Robert Allen apparently wrote to EPA administrator Douglas Costle, asking for a meeting if "he did not terminate the regional work." Some Idaho congressional staff reportedly offered to be present at Allen's proposed meeting with Costle, indicative of the high level support enjoyed by Bunker Hill in Idaho.[188]

The LIA was also challenging the standard in the courts and in 1980 asked the court to direct the EPA to reconsider the standard. The LIA claimed it had new information that a study conducted by Anthony Yankel, Ian von Lindern, and Stephen D. Walter contained errors that undermined the 1.5 µg/m^3 standard.[189] The study in question, "The Silver Valley Lead Study," examined the relationship between air and soil lead levels and children's blood lead concentrations near the Bunker Hill smelter.[190] The study was central to EPA's decision to set the ambient lead standard at 1.5 µg/m^3 and not higher. In 1980, as the LIA's challenge to the standard was making its way through the courts, Anthony Yankel, the lead author, refuted the study in an affidavit. He pointed to a "previously undetected error" that undermined its validity. Yankel asserted that if the error were corrected, the EPA standard would have been 3.75 µg/m^3.[191] Ian von Lindern was caught by surprise by his co-author's allegations and issued his own sworn statement in which he wrote that Yankel had recently visited him and admitted to working for Bunker Hill. Von Lindern felt there was no basis for Yankel's allegations.[192] The court rejected the Yankel/LIA challenge to the standard, and it was ultimately upheld.[193]

In the end Bunker Hill closed the smelter without meeting the ambient lead standard. The company's attempts to downplay the childhood lead poisoning problem in the Silver Valley must been seen in the context of the larger agenda of ILZRO and the LIA to downplay the problem of airborne lead and subclinical health effects in the hopes of influencing federal environmental and occupational lead standards. The Silver Valley community was a pawn in this higher stakes game.

The other major regulatory effort at the federal level was the OSHA lead standard that Bunker Hill and other lead smelters were also fighting. OSHA set a lead standard in 1978 that restricted lead in plant air to 50 µg/m^3. Bunker Hill was working with ASARCO and ILZRO to enlist the United Steelworkers of America (USW) support in contesting the OSHA standard. In 1979 Bunker Hill's in plant air lead levels averaged over 1,000 µg/m^3 and sometimes were as high as 17,000 µg/m^3. Most workers in smelter lead exposure areas had blood lead levels over 50 µg/dL and 42 percent had levels over 60 µg/dL.[194]

Of course some of the children poisoned in the Silver Valley were children of smelter workers, many of whom were also suffering from lead exposure and lead poisoning. At times Bunker Hill workers advocated for improvements in working conditions, and at times fears of job losses quieted these concerns. When news of the lead poisoning epidemic broke in 1974, at a public meeting a "Union representative for the United Steelworkers made a lengthy talk in which he combined concern for the health of the children with concern for the health of those who work in the plant as well as concern to make sure that the smelter kept operating so that employment would not be affected."[195] The USW Local representing Bunker Hill workers formally asked the federal government to conduct a health hazard evaluation at the smelter in August of 1975, exercising its right to be informed about workplace hazards.[196] At the federal level the USW fought hard to get OSHA to adopt a stringent standard for lead exposure in industry and criticized the agency for taking many years to do so. It also worked to expose abuses within the lead industry, such as the prophylactic use of chelating agents in workers.[197]

Bunker Hill workers, like other smelter workers, were in a precarious position in the 1970s and 1980s. Broader forces were shaping the future of the smelting industry in the United States, making many western smelters, in the words of historian Laurie Mercier, "an anachronism." Her analysis of the position of the working class in her study of Anaconda smelter workers in advocating for improvements in health and safety is relevant to Bunker Hill, where workers seemed to flip-flop between supporting and protesting OSHA involvement. Mercier writes: "In an unpredictable economy in which workers do not control the means of the production or have even minimal job protection, they will forgo their own health to keep jobs that provide the sustenance for their families and communities. The dominant position of capital often has led workers to spurn their old environmentalist allies on health and safety issues."[198]

In the summer of 1979 OSHA was stepping up pressure on the company to protect workers by requiring the use of respirators in some parts of the plant until engineering controls could drive down air lead levels in the workplace. A small number of Bunker Hill union members called a meeting to determine whether the union should oppose OSHA's enforcement of lead and arsenic standards at the smelter. Idaho senator James McClure was outspoken on the issue, alleging that regulation could lead to job losses and calling OSHA's decision to significantly lower worker's exposure to arsenic at the plant a "drastic action."[199] In this politicized environment where workers had deep anxieties about job losses, the union was walking a fine line in advocating for health and safety protections.

By the end of the 1970s real divisions also persisted in the community over the issue of whether children had been harmed by Bunker Hill's lead emissions. Although many Silver Valley families defended Bunker Hill and denied that their children were injured, the community was not of one mind. Some parents and community members questioned Bunker Hill and public health authorities and asserted that they saw real problems in their children or were worried about such problems. A family that had two lead poisoned children was the first to sue Bunker Hill, but the case was dropped before it went to court. The Yoss and Dennis families filed suit against Bunker Hill in 1977 for $20 million in damages. As the Yoss case was making its way to trial at the end of 1979, more parents were speaking publicly about problems their children were having, particularly some of those who were highly exposed or poisoned in 1974.

In an investigative piece written for the *Washington Post* in 1979, reporter Bill Richards set out to explore whether children had been harmed. He began making inquiries of Bunker Hill/Gulf managers and raising questions about the Shoshone Project and Bunker Hill's and Glen Wegner's roles in the project. Internally Bunker Hill managers were extremely concerned about how to handle these inquiries and feared a "disaster" and "an impending PR problem of major magnitude."[200]

The Richards article, which appeared in October of 1979, led with the sentence "Something terrible is happening to children here" and reported that the parents of more than thirty children interviewed in Kellogg and elsewhere believed that their children had been harmed by their exposure to heavy metals in Kellogg's air. One mother said her six-year-old daughter could not follow simple instructions. Another family's two sons were in special education classes. Said the father, "They can't seem to concentrate on anything or remember what you say to them."[201] At age fifteen, one of the boys was reportedly doing kindergarten-level work. Another, a nine-year old girl, reportedly spent three years in kindergarten and was unable to write the alphabet.[202]

Highly exposed children were struggling with multiple problems six years after peak exposure. In one case a physician's evaluation noted that it was "alarming" that a child who had moved away from the smelter years before still had elevated blood lead. He wrote that, despite very high blood lead levels, the child "appears to have never had adequate treatment with chelating agents."[203] Some children with very high blood lead levels also received chelation treatment as outpatients, even though the CDC recommended their hospitalization.[204] Chelation without intensive medical supervision was generally not recommended, and there was also some concern at the time that, among children exposed to cadmium (as Silver Valley children were), chelation might deposit cadmium in the kidneys, causing further damage.[205]

Another family that lived within a half-mile of the smelter in 1974 spoke out publicly about its harrowing experience with lead poisoning. When Gloria Dixon found out her children had elevated levels of lead in their blood, she said, "I didn't know what to do or where to turn. . . . I found out the lead in their blood came from the air they were breathing. I wanted to lock them up in the house so they might be safe." Her children were students at the Silver King school. She testified to a congressional committee that when she refused to send her children to school for fear of further exposure, she was threatened with jail. Several of her children experienced a range of health problems from lead exposure. She told the committee, "When my son falls down and skins his knee and cries, I kiss it and make it better; but what do I do when he wants to know why he has brain damage?"[206]

Other parents did not sue but had criticisms of how the state and Bunker Hill were handling the problem and made those concerns known. Andrea Cornish, who lived in Kingston, Idaho, wrote to Ed Gallagher and raised questions about toxic metals in the air and health problems suffered by residents of the Silver Valley. She also complained about the state relying on local physicians to monitor lead levels. In her case, she took her daughter, who was complaining of tiredness, nausea, and lethargy, to a local physician. The doctor did laboratory tests but not a lead test, which she found peculiar. Cornish told Gallagher, "It is difficult to be content with the gnawing fear that living here is damaging to your children."[207]

Silver Valley parents were clearly concerned when the lead poisoning crisis broke in 1974. When the state released the 1974 blood test results, between three hundred and four hundred people attended a public meeting to hear health officials discuss the results.[208] However, in the ensuing months and years the community would hear state health officials and the Shoshone Project officials downplay lead health concerns. This certainly complicated and confused the response of parents.

Even when health officials made clear, sound recommendations to the community, they might be questioned or attacked in the press. In May 1975 the health department recommended that people living in Kellogg, Smelterville, and Pinehurst should not eat locally grown vegetables due to high lead content.²⁰⁹ James Halley, president of Bunker Hill, was quoted in a news article discussing the recommendation as saying, "People are sick of the government telling us what is good for us."²¹⁰

Health officials discussed among themselves the community's attitude toward the crisis and disagreed as to whether the community was complacent or the public health community had "failed in accurately alerting the people to just what a serious problem we have had."²¹¹ Ian von Lindern characterized health department staff as believing the community's complacency was at least in part the fault of "softly worded news releases" coming from the Shoshone Project, which may have "lulled the community into a false sense of security."²¹² It is clear that the community got little detailed information about the health effects of lead from state health officials. The community was also counseled to use local doctors, some of whom had a financial relationship with Bunker Hill.²¹³

In 1981, when new questions about the smelter's effects on children's health were raised in connection with the upcoming trial alleging that children were damaged, Bob Dunn, a smelter worker, said, "I go by the [state] tests and they say my kids are all right. But I feel people who are really worried about it should get the hell out."²¹⁴ While there were clearly strong and differing opinions in the community, the contention that parents did not care about the problem or their children's health served Bunker's interests and probably limited investigations into health problems near the smelter.

The *Yoss* case was ultimately settled in 1982 for $8.8 million after a five-week trial and testimony from more than twenty doctors and neuropsychologists. The following year the company reached a $23 million settlement with an additional thirty-three children and four adults who claimed psychological and physical damage from lead pollution. The case broke new legal ground because it was the first time that victims of lead poisoning resulting from industrial pollution had ever won compensation for damages through the courts.²¹⁵ Although some families received compensation, no Gulf Resources or Bunker Hill officials were personally held accountable.

The resolution of the *Yoss* case came at a high cost to the families involved who found themselves the object of community ire. It did not help matters that the settlement coincided with Gulf's announcement that it would close the smelter for good. Some in the Silver Valley blamed the families who sued for the closure. In the press, one Kellogg resident called the families "dirty, uneducated, low-mentality white trash," and another said the kids were poisoned because they "didn't know how to use soap and water. It was their own darn fault, but they blamed Bunker Hill."²¹⁶ Even Idaho's state epidemiologist echoed the claim that the children's high blood lead levels were due to living in unhygienic conditions, and questioned why Gulf should have had to pay damages in the lawsuits.²¹⁷ Paul Whelan, one of lawyers for the plaintiffs, recalled that he and his partner were armed with a gun when they went to the courthouse to finalize the settlement. They had received death threats and feared their lives were at risk, given some sentiments in the community.[218]

The closure of the Bunker Hill mining and smelting operation was announced in August of 1981. In 1980 Bunker Hill provided Gulf $338 million in revenue, but after $8 million in losses in the first half of 1981 from depressed silver, lead, and zinc prices, Gulf decided to close its Bunker Hill operations.[219]

The Bunker Hill disaster, though extreme even by the standards of U.S. smelting communities, demonstrates how the lead industry sought to control the discourse over harm from airborne lead in smelting communities through research and public relations—even after an unprecedented community lead poisoning disaster was discovered—to protect the industry from liability but also to influence regulatory efforts. This story reinforces the importance of community vigilance, advocacy, and resistance to environmental pollution for galvanizing regulators to take action, and, in the absence of an organized and outspoken community, emphasizes how downplaying and minimizing serious pollution problems can have grave consequences for health and the environment.

5

Tacoma

A Disaster Is Discovered

Only two months after El Paso's lead poisoning crisis hit the front pages of newspapers across the United States, William Rodgers, a young professor of environmental law at the University of Washington, called for an investigation of the Tacoma smelter's impact on the public health of children living nearby. In a letter to the mayor of Tacoma and the chairman of the newly formed Puget Sound Air Pollution Control Agency (PSAPCA), Rodgers pointed out that the Tacoma smelter had much in common with the El Paso smelter; therefore, it was a matter of prudent public policy to study local children's exposure to toxic metals. He wrote, "More than fifty children [in El Paso] have been hospitalized with high lead levels in the blood, many of them on the recommendation of the company. The City has taken the extraordinary step of urging the removal of families from the vicinity of the stack." Rodgers called on government officials to act quickly to determine "precisely what amounts of lead and arsenic have been lost to the environment in Tacoma," to determine which agencies (local, state, or federal) could test for lead exposure, and to determine "whether company records on possible in-plant lead or arsenic poisoning of employees are available to cast light on a broader community health problem." His letter ended on a note of frustration: "After making several quiet telephone calls, I have found the response to this problem to be listless as usual. I hope the Board can get some action by looking into this matter and I commend you if you have already done so."[1]

Rodgers's frustrations were understandable. Although regulators and health officials at all levels of government had been aware since at least the early 1960s of the potential for significant public health and environmental problems from heavy metal exposure in the vicinity of the Tacoma smelter, no health investigations in the surrounding community had been conducted. The regional air pollution control agency, PSAPCA, was focusing its regulatory efforts directed at the Tacoma smelter almost entirely on sulfur dioxide emissions. This was in part reasonable. The sulfur dioxide problem was the most obvious, objectionable, and well understood. Little was known about the human health consequences of arsenic exposure among people living near nonferrous smelters. But suspicions that the plant's arsenic emissions were damaging the environment were long-standing. Locals said that there were no bees in the area and children were never stung because all of the bees were dead, killed by the smelter's toxic arsenic.[2]

At Tacoma, arsenic exposure rather than lead was the defining feature of the health crisis that emerged there shortly after the discovery of the lead poisoning epidemic in El Paso. Although the exposure of concern was different, what Tacoma shared with El Paso and Bunker Hill was that toxic metals were being emitted from the stack and ground-level sources in staggering amounts, and were detected in the bodies of people who lived there.[3] Toxic metals were also contaminating air, soil, house dust, and garden vegetables.[4] But there were other similarities as well. As Bunker Hill did in Idaho and ASARCO did in El Paso, ASARCO also sought to downplay concerns in Tacoma over the effects of smelter emissions on community health and to question the significance of the exposure to toxic metals in the community.[5] In some ways this was easier in Tacoma because of uncertainty about the specific health effects of community exposure to inhaled arsenic.

But based on what was known about arsenic from other exposure contexts, there was ample reason for a high level of concern for community health from arsenic exposure in Tacoma in the early 1970s. The State Health Department was concerned and so were some local health leaders, but this sentiment was not shared by all in the community; Tacomans were preoccupied with job losses and the region's struggling economy. It was also not shared during the 1970s by decision makers at EPA, the agency with the statutory authority to regulate arsenic in the air people were breathing,

Much of the public discussion between 1970 and smelter's closure in 1985 focused on uncertainty about possible health effects of exposure to arsenic and emphasized the lack of scientific evidence that people, particularly children, were being harmed. Industry proponents argued, implicitly and explicitly, that because health effects of arsenic exposure near the smelter were uncertain, the company should not be regulated to the extent that it would affect jobs. But as the decade of the 1970s unfolded environmentalists, local physicians, and some parents increasingly questioned this premise, essentially asking if health effects are so uncertain, why not take a precautionary approach to protecting children's health and force arsenic emissions to the lowest possible level?

In Tacoma, despite divisiveness in the community over regulating the smelter, there was some sustained community pressure on the company and on local regulators to reduce emissions. Although ASARCO was a significant local employer, Tacoma is part of a major metropolitan area, and the company did not have a stranglehold on jobs or public opinion. There was a range of perspectives on health risks among residents, an engaged environmentalist community concerned about regional environmental impacts, significant citizen organizing, and strong local legal and scientific expertise. These factors likely had much to do with the responsiveness of local regulators and the State Health Department in the 1970s.

A great deal of independent research was conducted on the smelter's effect on the environment during the 1970s, made possible by increased public spending for environmental research, and none of the studies allayed concerns about widespread impacts on the environment and the region's ecology. The findings from these studies, however, were not enough to prompt the EPA to take a strong stand on controlling the smelter's pollution. Reductions in emissions that did occur were incremental, and came primarily from local regulatory efforts.

As in El Paso and Bunker Hill, there were larger national policy considerations over the health effects of arsenic, its carcinogenicity, and regulating arsenic in air. While smelters were the most significant source of airborne arsenic, the carcinogen was also emitted by the mining sector, coal-fired power plants, the wood preservative industry, chemical production, glass manufacture, and waste incineration, giving it a powerful industrial constituency.

Although local regulators and officials from the EPA's regional office in Seattle urged action on the Tacoma smelter, at EPA headquarters in Washington, DC, regulation of this carcinogenic threat became stuck in a quagmire of delays. The EPA was requiring convincing evidence of community health effects before taking regulatory action on arsenic and seemed much more concerned about the possible economic consequences of regulation than the public health and environmental benefits of reducing exposure to a carcinogen.

A complicating factor for the EPA was that growing calls to regulate the Tacoma smelter occurred at a time when many smokestack industries were closing in the United States and blue-collar job losses were mounting. Despite public enthusiasm for clean air in the early 1970s, by the end of the 1970s the EPA appeared to be concerned about feeding a perception fostered by industry and others that environmental regulation cost jobs. Federal regulatory inaction in Tacoma in the 1970s and 1980s meant that the community had little choice but to turn to the courts to try to get redress. The Tacoma smelter continued to operate for fifteen years after the passage of the 1970 Clean Air Act, spewing arsenic all the while. Tacoma illustrates the long and difficult road for communities that have tried to abate toxic pollution even when there were laws on the books, and even when the exposure was to a known carcinogen.

The struggle to regulate Tacoma's arsenic emissions shares much in common with other so-called fenceline communities, those located close to polluting industrial facilities. In the absence of definitive information on the health risks of industrial emissions, strong regulations, and enforcement, such communities may languish for years in a regulatory no-man's-land, facing seemingly insurmountable barriers to achieving protection from pollution.[6] The Tacoma case illustrates the ambivalent response of government agencies whose staffs may be concerned about community health, but real or perceived political pressures on the agencies prevent or prolong decisive action.

Since much of the regulatory foot-dragging was blamed on scientific uncertainty about arsenic as a carcinogen, background on the medical and epidemiologic evidence for arsenic's carcinogenicity provides important context to the Tacoma controversy. Although significant histories of arsenic are available, these sources do not provide a detailed examination of the evolving medical and epidemiologic evidence for arsenic as a carcinogen in the postwar period.[7]

By the 1970s there was a long trail of inquiry into the effects of arsenic on human health, much of it concerning arsenic's acute toxicity. In the nineteenth and early twentieth centuries acute poisoning was often suspected when arsenic was found in the body and the victim suffered from classic symptoms such as profuse vomiting, stomach cramps, diarrhea, and the feeling of pins and needles in hands and feet. But it was much more difficult for physicians to make the link between lower-level chronic exposure to arsenic and adverse health effects, such as cancer, because years or decades could elapse between arsenic exposure and subsequent cancer. However, quite often people who are chronically exposed to low levels of arsenic experience classic symptoms of chronic arsenic poisoning (arsenicism), which include skin changes such as hyperpigmentation and thickening of the skin, black warty growths called keratoses, peripheral neuropathy, and sometimes perforation of the nasal septum. Dermatologists were among the first physicians to connect symptoms of chronic arsenic poisoning with subsequent cancer.

Arsenic was widely used in medicine to treat a range of illnesses until after World War II, particularly dermatological conditions. Popular treatments containing arsenic were Fowler's Solution, prescribed for skin conditions such as psoriasis; Donovan's Solution, and Asiatic pills.[8] Sir Jonathan Hutchinson, a British dermatologist, surgeon, ophthalmologist, and neurologist put forward the hypothesis that medicinal arsenic could cause skin cancer in 1887 after observing several cases in patients treated with arsenic.[9]

After Hutchinson's initial observation, dermatologists in England, Continental Europe, and the United States published numerous case reports of skin cancers among people who were treated with arsenical medicines. Unsuspecting patients who had taken Fowler's Solution for psoriasis or dry skin or Asiatic pills to improve their complexions wound up years later with black, warty, cancerous growths on their skin, which sometimes led to amputations or death. Case reports were often complete with gruesome photos of patients with cancerous growths on their hands, feet, backs, and elsewhere.[10]

By the 1930s some American dermatologists and public health officials were expressing strong concern over the ubiquity of exposure to arsenic in medicine and in the environment. Some thought arsenical skin cancer was underdiagnosed and had the potential to become a serious public health problem, given the widespread use of arsenic in medicine and agriculture, and in and around people's homes as an insecticide. Despite an apparent consensus in the literature by the 1930s that arsenic could cause skin cancer, some physicians continued to rely on it to treat illness, and its use continued in agriculture.[11]

Some of the early case reports of skin cancer among people exposed to arsenic were agricultural and smelter workers, but exposure to arsenic among workers did not arouse much concern in the industrial hygiene community on either side of the Atlantic in the early to mid-twentieth century. In the workplace, arsenic was considered an acute poison but not one contributing to chronic health effects. For example, in 1916 British occupational physician Thomas Legge wrote that exposure to arsenic did not produce "a slow undermining of the health as is commonly caused by lead poisoning."[12] The prevailing view among early-twentieth-century occupational hygienists in the United States was that the chief hazards of occupational exposure to arsenic exposure were irritant effects on mucous membranes and in some cases peripheral neuritis. There was little concern about systemic effects from exposure on the job or the effects of long-term exposure, including cancer.[13]

Wilhelm Hueper took exception to this view and strongly criticized the lack of concern in the U.S. public health and occupational hygiene community regarding worker exposure to arsenic. Writing in 1942, he said, "The impression prevailed even up to very recent times among some public health officials of this country that chronic arsenico-dermia of industrial origin did not exist in the United States." With obvious incredulity, Hueper pointed to the "extraordinary immunity of American workers to arsenical malignancy" and suggested that industrial arsenicism was not being recognized by physicians.[14]

In fact, upon combing through all published case reports of arsenical skin cancer attributable to occupational exposure, Heuper found that the United States was the country with the most case reports of arsenical skin cancer and arsenic dermatoses in workers in the ten years prior to 1942. He argued that this only made sense given the way arsenicals were handled in agriculture: "It is scarcely possible to take less precautionary measures and to handle arsenicals with more utter disregard of the general health than that which has been practiced in our country in recent years. Anyone who has watched the dusters and sprayers of arsenical insecticides at work must have been impressed by the almost supreme carelessness with which these poisonous substances are dispensed."[15]

Skin cancer was not the only cancer that nineteenth-century physicians thought might be related to occupational arsenic exposure. In the early 1820s British physician John Ayrton Paris had observed that scrotal cancer among smelter workers might be due to arsenic exposure. A century later some French physicians thought arsenic played a significant role in many occupational cancers and specifically might cause bladder cancer in aniline dye workers. By the 1940s there was increased attention to the possibility that arsenic might cause lung cancer in exposed workers. The high mortality rate from lung cancer among miners in Schneeberg, Germany, and Joachimsthal, Czechoslovakia, led early researchers in 1879 and some subsequent investigators to point to arsenic as a potential causative agent. In 1945 the British Factory Department commissioned a review of the worldwide literature on arsenic carcinogenesis due to long-standing concerns. The British had been tracking industrial reports of lung cancer associated with arsenic since 1939.[16]

The British review, published in 1947, viewed as a settled matter that medicinal use of arsenic could cause skin cancer. Additionally, occupational exposure and exposure through drinking water were also linked to skin cancer. Occupational cases, however, were thought to be relatively rare. The evidence for internal cancers related to arsenic was not considered very strong, in part due to idiosyncratic and inconsistent findings from different parts of the world and different exposure contexts. For example, in reference to occupational cancers, different cancer sites were noted in different occupational settings (e.g., cancer of the scrotum in workers exposed to arsenic in by-products of coal, bladder cancer in aniline dye workers,[17] skin cancers in users of arsenical pesticides, and lung cancers in miners), a finding that seemed implausible.[18]

A few years later, however, what has been called the first historical occupational cohort study, conducted by A. Bradford Hill and E. Lewis Faning, provided early evidence of the relationship between arsenic and lung cancer in arsenic-exposed workers. Hill and Faning studied worker mortality at a sheep-dip manufacturing factory in England. Arsenic-exposed workers were found to have higher lung cancer mortality than other workers in the same town. Other evidence for arsenic's role in lung cancer across different industries was published in the ensuing years. Vintners in the Moselle Valley of Germany who used arsenic as a pesticide on wine grapes were found at autopsy to have cancers of various organs, including the lungs. In 1957 miners in Rhodesia were reported to have higher than expected rates of lung cancer, which the researcher hypothesized was related to exposure to arsenic-bearing gold ore.[19]

In the United States up until the postwar period, little research had been done on arsenic and its role in causing occupational cancer, although U.S. researchers were aware of the work of their colleagues overseas. But after the Hill and Faning study showed increased respiratory cancer mortality in arsenic-exposed workers, ASARCO began to take an active role in occupational arsenic research and funded Harvard researchers to conduct the first U.S. study of cancer in smelter workers exposed to arsenic. The study compared cancer deaths at the Tacoma smelter, where workers were highly exposed to arsenic, to another smelter where workers were unexposed. In the paper published in 1951 the researchers did not acknowledge that ASARCO had provided the funding for the study nor that Tacoma was the high-arsenic smelter. The authors concluded that "the handling of arsenic trioxide in the industry studied does not produce a significant change in the cancer mortality of the plant employees."[20]

By the end of the 1950s ASARCO had grown even more interested in research on arsenic's carcinogenic properties. Congress was considering the Delaney Clause, which would prohibit known carcinogens as residue on food.[21] At this time ASARCO was producing about one thousand tons of arsenic per month at Tacoma alone, much of which was used in agriculture. If its use on food crops was restricted or prohibited under the Delaney Clause, the company would have a limited market for this toxic smelting by-product. This was a significant concern since storage was expensive and disposal fraught with difficulty and liability.

Concern in the late 1950s about possible restrictions on arsenic residue on food prompted meetings of arsenic industry representatives, including ASARCO and the National Agricultural Chemicals Association, to "present a united front" in addressing questions about its carcinogenicity.[22] To aid industry efforts, the ASARCO medical staff was asked to conduct further study of Tacoma smelter workers and their cancer risk.[23] The request resulted in a mortality study published in 1963 that concluded: "Arsenic trioxide absorption of the range we have described in industry does not cause cancer."[24] The lead author of the study was Tacoma smelter physician Sherman Pinto.

Although in Europe the case for arsenic as a skin carcinogen seemed to be unassailable by the end of World War II[25] and the Hill and Faning study made its role in causing lung cancer seem plausible, in the United States there were few alarm bells sounding about arsenic's carcinogenic potential, particularly in occupational settings. For example, Bert L. Vallee, a Harvard University Medical School researcher, wrote an influential review of arsenic toxicology in 1960 that stated, "Analysis of the cases reported to date leads to the conclusion that arsenic may be a factor in predisposition to skin cancer but at best in a very small per cent of persons exposed and those mainly by way of medicinal drugs and rarely in industry. There is no valid basis for assuming that the element plays any role in causing other types of carcinoma."[26]

But other studies of medicinal, workplace, and environmental exposure published in the 1960s reinforced concerns about arsenic's carcinogenicity. British investigators in 1963 reported on six cases of lung cancer, three of which were in nonsmokers, in people who had previously been treated with arsenical medications and had experienced classic skin manifestations of arsenic poisoning. Another 1963 study, this one by the National Cancer Institute (NCI) on the mortality in underground metal miners, found an increased risk of lung cancer that the authors speculated could be related to arsenic. In 1969 NCI published the first independent epidemiologic study of U.S. smelter workers to examine their cancer risk. The study of the Anaconda copper smelter in Montana found a three- to eightfold increased risk of lung cancer among that plant's arsenic-exposed workers. Also in the 1960s skin cancers were reported in people in Taiwan in an area with high concentrations of arsenic in ground water, consistent with earlier reports from Argentina and Poland.[27] By the early 1970s evidence from case reports, case series, and some epidemiologic studies of workplace and environmental exposure pointed to arsenic as a skin and lung carcinogen.

Despite the accumulating evidence for arsenic's carcinogenicity in the early 1970s, arsenic has proven to be harder to study and more difficult to monitor and address as a public health threat in community settings than lead. Although the capacity to detect arsenic in organic samples in very small quantities dates back to the 1830s, interpreting test results remains challenging even today.[28] Detection in hair and nails reflects chronic exposure, and arsenic in blood indicates very recent exposure, within the last few hours. However, arsenic is most typically measured in urine, and urine arsenic concentrations are used in clinical medicine and in epidemiologic studies to measure exposure. Detection of arsenic in urine indicates that exposure occurred within the past several days. Urinary arsenic, however, is an imperfect biomarker of exposure due to the fact that organic arsenic, which is found in seafood and is thought to be harmless, can cause high urinary arsenic concentrations. The inorganic species are highly toxic, but few laboratories have the ability to speciate, or separate, inorganic from organic arsenic species to determine the contribution of each to elevated urinary arsenic. An elevated urinary arsenic concentration could be relatively harmless and caused by eating a fish dinner the night before rather than exposure to carcinogenic inorganic arsenic.[29]

The medical and public health communities have not set public health guidelines for arsenic in urine, in part due to the difficulty of interpreting urinary arsenic because of the contribution of organic species from seafood. This has also likely hindered research into the health effects of inorganic arsenic in the United States. Research into arsenic and its impact on children's health in particular lags behind research on lead, despite significant recent advances. While many researchers are very concerned about the health effects of inorganic arsenic, in the United States arsenic lacks the organized medical and public health constituency that has effectively driven down exposure to lead among children in the past several decades.

In addition to lacking public health guidelines for arsenic exposure, there is no agreement in the public health literature on what would constitute a "normal" level of urinary arsenic in U.S. populations as a less formal standard to determine when one's exposure to arsenic is abnormal or of concern. Because arsenic is ubiquitous in the environment, most people will have arsenic in their urine at any given time. Adding to the confusion in the public health literature, the concept of "normal urinary arsenic" has been used to refer both to the average concentration in people with no exceptional exposures and to the concentration that will not cause harm.

Attempts to define "normal" arsenic content in urine and in other bodily tissue and fluids occur go back more than a century, when early investigators frequently found arsenic in the urine and other bodily fluids of those tested and tried to determine whether this was because arsenic was a normal constituent of the body or if its occurrence signified low-level poisoning or intentional exposure.[30] In the late 1930s and early 1940s, because of persistent concerns about arsenic residue on sprayed fruits and vegetables, both the U.S. Public Health Service (USPHS) and industry took up this issue of "normal" urinary arsenic levels. A study of urinary arsenic concentrations in people with no known exposure was done by the USPHS in 1941 among families affiliated with the National Institute of Health in Bethesda, Maryland. Of forty-six participants, all but four had measureable amounts of urinary arsenic, with an average concentration of 14 µg/L.[31]

International concerns about arsenic and lead residue on U.S. fruit and vegetables led Congress to direct the USPHS to study Washington State fruit growers and residents in and around Wenatchee, where large quantities of arsenical pesticides were used. In the late 1930s Wenatchee area growers were using upward of 7 million pounds per year of lead arsenate on orchard crops. The study measured lead and arsenic in blood and urine in orchardists and residents, including children. The research largely exonerated lead arsenate from causing any health effects, did not consider subclinical disease or chronic health effects, and gave short shrift to arsenic as a carcinogen.[32] The overt implication of the Wenatchee study was that since those most exposed to lead arsenate were not suffering obvious health effects, neither would consumers of fruit, who were exposed at lower levels. The Wenatchee study was used to justify raising tolerances for lead and arsenic on fruit.[33] The study also led some to argue that relatively high concentrations of arsenic in urine could be "normal"—that is, without health effects.[34]

Defining the level of urinary arsenic that was "normal" and the level associated with harm was central to understanding arsenic's harmful effects and determining who might be at risk. The arsenic industry had a stake in weighing in on these questions, and it did so by publishing studies in the peer-reviewed literature. In 1953 ASARCO reported that "normal individuals not exposed to arsenic trioxide dust" who participated in a study had average urinary arsenic concentrations of 130 μg/L. Tacoma smelter workers who were exposed to arsenic had average urinary arsenic concentration of 820 μg/L, and ASARCO researchers reported that these concentrations were not associated with systemic illness. They also concluded that concentrations in urine up to 4,000 to 5,000 μg/L were not necessarily harmful, and that "much more arsenic can go through the human body without causing sickness than has heretofore been realized."[35]

In 1958 researchers from the Industrial Hygiene Foundation (an industry-sponsored research foundation) wrote that in those with "no unusual exposure," arsenic concentrations are typically less than or equal to 100 μg/L but stated that "high urinary arsenic levels may occur with no evidence or indications of any adverse effects."[36] ASARCO's research on urinary arsenic concentrations in its workers would later be relied upon by the USPHS in its response to a community arsenic poisoning episode among children living in a Nevada mining camp who were exposed to arsenic dust. They had skin irritations from arsenic but since their urinary arsenic concentrations were similar to those found in ASARCO workers (820 μg/L) and reported to be "benign," the investigators concluded that their exposure was not excessive and their dermatologic problems were not caused by systemic poisoning.[37]

Even by the early 1970s there was still no consensus on "normal" urinary arsenic; in fact, the third edition of Alice Hamilton and Harriet Hardy's *Industrial Toxicology* text in 1974 called the literature on urinary arsenic concentrations "confusing" but reported that normal levels of arsenic in the urine range from 13 to 46 μg/L.[38]

Today there is still disagreement over the level at which urinary arsenic can be considered normal. According to a recent Agency for Toxic Substances and Disease Registry (ATSDR) publication, normal urinary arsenic is less than 100 μg/L.[39] Philip Landrigan reviewed the literature in the early 1980s and put "normal" at less than 50 μg/L.[40] In a recent representative sample of U.S. residents, mean total urinary arsenic concentrations ranged from 7 to 8.5 μg/L in six- to eleven-year-olds, teenagers, and adults.[41]

The lack of public health guidelines for arsenic in urine and disagreement over "normal" complicated the interpretation of the risks faced by children in Ruston and Tacoma who were found to be excreting arsenic in their urine. Added to this, there were few studies correlating specific urinary arsenic concentrations with specific health effects. Without clear public health guidelines, concerns about low-level arsenic exposure were much more contestable than concerns about low-level lead.

In spite of the unknowns, by the 1970s, except among some skeptics, arsenic was an acknowledged skin carcinogen, and evidence was growing for its role in lung cancer. Less clear were the chronic effects of breathing arsenic in air, particularly for children, especially for noncancer endpoints, which were arguably more immediately relevant to children.

This was the scientific context when William Rodgers began making "quiet telephone calls" in May of 1972 to learn what was known about the health impacts of the Tacoma smelter's emissions. Coinciding with Rodgers's inquiries, Dr. Sam Milham, a Washington State Department of Health epidemiologist, was asking federal scientists how best to study lead and arsenic exposure in children living near the Tacoma smelter.[42]

The red-brick Ruston Elementary School sat in the shadow of the smelter's smokestack, a stone's throw from the property boundary. Sam Milham recognized the school as the most likely epicenter of children's exposure to toxic metals. In 1972 he began taking samples of Ruston schoolchildren's blood, urine, and hair for the purposes of measuring lead and arsenic exposure. For comparison, Milham also took similar measurements from children who attended Fern Hill Elementary School, about eight miles to the southeast. By the fall of 1972 Milham's sampling had uncovered some alarming evidence of community arsenic exposure. Seven children living within a half-mile of the smelter stack were found to have average urinary arsenic concentrations of 300 µg/L—a level comparable to some workers in the smelter.[43] Between one-half mile and one mile of the stack, of eight children sampled, the average urinary arsenic concentration was 190 µg/L. Samples taken from nine Ruston preschool children revealed a mean urinary arsenic concentration of 270 µg/L. The highest urinary arsenic concentration measured in early sampling was 620 µg/L in a young child living near the stack.[44] Children's urinary arsenic concentrations showed a pattern of decreasing with distance from the stack, indicating that the smelter was the source. High concentrations of arsenic were also found in sampled vacuum cleaner dust, which confirmed that the interiors of houses near the smelter were also contaminated.[45]

The results were made public in mid-November 1972, with health officials calling the arsenic concentrations found in children living near the plant more than fifteen times the "safe level."[46] Dr. Wallace Lane, the assistant secretary of the Washington State Department of Health, walked a fine line, trying not to alarm parents but reinforcing the concern of state health officials: "While not in imminent danger of acute arsenic poisoning, the health of Tacoma and Ruston residents is affected by exposure to arsenic at the levels we are finding."[47]

FIGURE 5.1 Dr. Samuel Milham of the Washington State Department of Health in Ruston with the Tacoma smelter stack in background, 1972.
National Archives and Records Administration, RG 412,
Environmental Protection Agency. Photographer: Gene Daniels.

Although the smelter also emitted lead, early on state health officials dismissed the Tacoma smelter's contribution to children's elevated blood lead in the region. Since the level of concern at the time was 40 μg/dL, the blood lead levels (BLL) found in Ruston third and fourth graders were lower than this cutoff (ranging from 8 to 30 μg/dL), although some younger children living closer to the smelter were found to have BLLs near the level of concern, with a mean of 36 μg/dL.[48] But the state never conducted a thorough investigation of Ruston children's lead exposure, and by 1974, on the basis of the limited lead testing in 1972 (approximately twenty-seven samples from Ruston schoolchildren and eight preschoolers), state health officials had concluded that "lead contamination is not a problem in this area."[49] Mercury and cadmium exposure were also not investigated in an in-depth manner in the community, although these heavy metals were also in found in smelter emissions.

Because there was no public health standard for arsenic in urine, Sam Milham thought that the best population to compare Ruston children's urinary arsenic levels to were the Bethesda, Maryland, children, who composed part of the control community in the USPHS Wenatchee study from 1941.[50] The children apparently had no known exposure to arsenic, although arsenic was certainly in the food supply due to its use in agriculture. To Milham, the average concentration of 14 μg/L found in the Bethesda children represented an appropriate definition of "normal," or background urinary arsenic.

Because he had found urinary arsenic concentrations in some Ruston children that were over fifteen times higher than this, he turned to the public health and medical literature to understand whether health effects had been associated with this level of exposure in children. But there were few studies available that could shed light on what the children's urinary arsenic concentrations might mean to their immediate and long-term health. What seemed most relevant were Dr. Sherman Pinto's two studies of Tacoma smelter workers because Pinto had reported on morbidity and mortality in workers at various concentrations of urinary arsenic. And even though both of Pinto's studies had concluded that worker health was not significantly compromised by arsenic exposure, a closer look at the data showed that workers with arsenic concentrations in the range of Ruston's children suffered important health effects. The mortality data pointed to increased respiratory cancer mortality in both "unexposed" (130 µg/L) (suggesting that these workers were inappropriately categorized as unexposed) and "exposed" (820 µg/L) workers.[51] Arsenical dermatoses were also evident in the majority of workers with mean urinary arsenic concentrations of 600 µg/L.[52]

Although Sherman Pinto, the Tacoma smelter physician, had not attributed the increased respiratory cancer mortality evident in his 1963 study to arsenic, Sam Milham considered this the most likely explanation, since the NCI's 1969 study of arsenic-exposed Anaconda smelter workers had reported a three- to eightfold increased risk of respiratory cancer mortality.[53]

An interesting aside is that ASARCO scientists were very concerned about the NCI's finding of increased lung cancer death in smelter workers. Somehow they received a copy of the Anaconda study prior to its publication and discussed internally their intention to influence the study's findings, its publication, or both because, as ASARCO's vice president for environmental affairs, Kenneth Nelson, wrote, "We don't need any more trouble than we have with fears about asbestos, cadmium, and lead."[54]

Dr. Wallace Lane, assistant secretary of the Washington State Department of Health in 1972, was alarmed that some urinary arsenic concentrations found in Ruston children were within the range that both cancer and noncancer effects were found in Tacoma smelter workers. In November of that year he requested that PSAPCA develop and implement an ambient air standard for arsenic. Lane wrote that there was "an urgent need" for adoption and enforcement of a standard. PSAPCA agreed, and requested the EPA's help and technical assistance with studies that "may lead to appropriate ambient air and emission standards for arsenic in our region."[55]

Of course, ASARCO was not sitting still with all of the new investigatory focus on Tacoma. As in other smelting communities as concerns about the industry's impact on the environment and public health grew, public relations efforts began to figure prominently in ASARCO's response to growing challenges in Tacoma. Even before public concern became fixed on arsenic, Simon D. Strauss, then ASARCO's executive vice president, directed company officials to start an "intensive public relations campaign on behalf of the Tacoma smelter" in early 1972 to fight local calls for strict limits on sulfur dioxide emissions. The first objective was to "replace the image in Tacoma, now considered to be 'don't give a damn,' with the image of a good corporate citizen," and others were to "impress upon" Puget Sound region voters "the reasonableness of its [ASARCO's] position on air quality standards" and to convince PSAPCA to accept EPA's sulfur dioxide standard rather than PSAPCA's more stringent standard for control. The company's public relations efforts would subsequently focus on the issue of arsenic in air as well. Strauss thought the regulatory objectives could be achieved by getting publicity for ASARCO's "environmental progress," sponsoring community activities, cultivating influential citizens in Seattle and Tacoma, and "other imaginative means of influencing various publics."[56]

Although the company had announced its intention to achieve just 50 percent control of its sulfur dioxide emissions even though a new PSAPCA regulation called for 90 percent control, the *Tacoma News Tribune* touted the company's commitment to pollution control in a July 1972 editorial. The editorialist noted that the company's management had in the past year "developed a more active interest in becoming a citizen of the community." Planned pollution controls were an indication that "ASARCO, having determined it will stay in Tacoma, has decided also to do it right."[57]

But the company's public face contrasted with its behind-the-scenes efforts to protect itself from growing public concern. In March of 1972, before the state's investigation into children's exposure to arsenic began, company officials were collecting and analyzing their own data that confirmed a serious environmental contamination problem in the area. Following the El Paso discoveries, ASARCO began soil testing in Ruston on vacant lots and in children's play areas. The testing showed that Ruston children were playing in highly contaminated areas. Soil lead levels ranged from 70 to 4,800 parts per million, and arsenic concentrations ranged from 46 to 3,302 parts per million.[58] Natural background concentrations of lead and arsenic in western Washington soils are 24 and 7 parts per million, respectively.

But ASARCO apparently took no action to notify the community about the high levels of lead and arsenic found in its sampling. The company's Western Smelting director, L. C. Travis, wrote internally, "Samples from six areas are high enough to cause me some concern but I don't know what to do about it. It seems to me we would stir up a lot of bad publicity if we tried to replace the soil like we did at El Paso and I am inclined to let the 'sleeping dog lie' unless some Agency calls the matter to our attention."[59]

In the spring of 1972, aware of the state's planning for heavy metal testing in Ruston children, ASARCO officials were making contact with local health officials to determine the types of health studies that were being conducted or planned, and were offering the company's cooperation.[60] Offering to cooperate seemed to aim at increasing ASARCO's chances of having its own medical experts involved in any planned studies and avoiding being surprised by the findings rather than a genuine desire to collaborate with local government officials on matters concerning public health since just a few months later company officials were denying access to Tacoma smelter worker's medical records by the State Health Department.[61] The health department thought that worker medical records would hold clues to the kinds of health problems that might be found in the community and would provide a basis for developing a community health investigation.

In keeping with its stance in El Paso, ASARCO argued to the State Health Department that elevated urinary arsenic concentrations in Ruston children were attributable to ingestion rather than inhalation of contaminated air.[62] A finding that inhalation was the primary route of exposure would lend support to calls to either capture more of the plant's arsenic emissions or shut the smelter down.

In public statements the company cast doubt on arsenic's potential to damage the public's health. For example, at a 1973 hearing in Tacoma ASARCO called on its expert witness, the Washington State toxicologist, who said arsenic "is certainly not a strong carcinogenic agent."[63] ASARCO scientists and experts frequently pointed to the fact that "arsenic has not been shown to be carcinogenic in animal experimentation."[64] Tacomans were routinely assured that there was no evidence that community health was being harmed from the plant's arsenic emissions.[65]

Coinciding with the investigation of children's exposure to arsenic in Tacoma, a large number of environmental health investigations took place in the Puget Sound region in the 1970s in an attempt to quantify the environmental impact of the smelter. These investigations demonstrated that inorganic arsenic was dispersed widely in Ruston, Tacoma, and beyond. Scientists tested soils, vegetation, and sediments in and around North Tacoma and across the Dalco Strait on Vashon Island. They found arsenic at concentrations over background nearly everywhere they looked. By the mid-1970s a growing body of studies had identified a regional contamination problem. Arsenic pollution was found in soils, air, vegetation, and even lake sediments in parts of Lake Washington, more than twenty miles to the northeast. Even at this point, some researchers thought heavy metal contamination extended to Seattle, thirty miles northeast, or beyond.[66] The studies in Tacoma were made possible by an expanding federal investment in independent environmental research contributing to a burgeoning environmental science that looked in novel ways at the environmental fate of pollutants. In Tacoma such studies investigated atmospheric transport of pollution, the impact of arsenic and sulfur dioxide on rain chemistry, deposition of heavy metals in soils and sediments, and their effects on marine life, vegetation, and domestic animals, among other things.

For public health researchers and environmental regulators, one of the most important questions to answer was how much arsenic was in the air that people in Tacoma were breathing. In 1972 regulatory agencies knew very little about ambient air concentrations of arsenic near the smelter since they had no continuous monitoring program in place. There were, however, some small studies done in the early 1970s that provided evidence that air concentrations were high—higher than anywhere else in the United States. Government air sampling from the early 1970s turned up average arsenic concentrations of 1.5 µg/m^3 two miles from the smelter, with a maximum concentration of 9.8 µg/m^3, eighty-nine times higher than the U.S. urban average. Even in the area beyond two miles from the smelter, the maximum concentration was close to 2.0 µg/m^3. The arsenic concentrations measured in Tacoma were orders of magnitude higher even than those found near other smelters.[67]

By 1974 the EPA had developed estimates of the smelter's tall-stack emissions, which put arsenic at 58 pounds per hour (250 tons per year), lead at 24.6 pounds per hour (106 tons per year), and cadmium at 1.3 pounds per hour (5.8 tons per year).[68] At this time, however, little was known about the plant's fugitive or low-level emissions—those that did not go through the stacks. PSAPCA estimated that fugitive particulate emissions in 1973 were 1,640 tons, but the issue had not been studied in depth, even though both the EPA and PSAPCA recognized that fugitive emissions could have a more significant impact on local ambient air concentrations than stack emissions, which could be transported far and wide due to the tall stack.[69] Sampling of particulates in the vicinity of the smelter found that 40 to 70 percent of the particulates were of an "optimum" size for "alveolar deposition."[70] In other words, a high percentage of the toxic metals emitted by the smelter were perfectly sized for lodging in the lungs of local residents.

None of the early 1970s environmental monitoring studies allayed concerns about the impact of the smelter. Each added another dimension to what appeared to be a pervasive problem of environmental contamination from ongoing and historical air emissions. Smelter emissions even had a dramatic effect on shaping the landscape of the area. An EPA-sponsored study found that within a mile to the south-southwest of the smelter, the area "is striking in that only a few species of vegetation remain with a complete absence of legumes (alfalfa, clover, etc.) and Douglas fir." According to the study, even four to five miles to the southwest Douglas fir trees, an iconic Northwest tree, did not grow.[71]

What was notable about the Health Department and PSAPCA response to the Tacoma smelter's pollution in the early 1970s was the relative rapidity with which the children's exposure and environmental monitoring studies were conducted. In a few short years a great deal of data was collected that signaled a regional pollution and human exposure problem. However, left unanswered by the public health and environmental monitoring studies was what would be done about the smelter and its considerable pollution. PSAPCA and the health department were in agreement that community arsenic exposure should be abated, but this would prove to be much more complicated than anyone anticipated in the early 1970s.

6

A Carcinogenic Threat

If you drive north on Pearl Street in Tacoma and cross Forty-ninth Street heading toward Puget Sound, Pearl Street marks the division between the town of Ruston, on your right, and the city of Tacoma, on your left. Weathered wooden houses line Pearl Street and are interspersed with some local favorite businesses—the Antique Sandwich Company in Tacoma and the Ruston Inn on the other side of the street. You can signal your municipal allegiance by where you choose to have lunch.[1]

Continue down Pearl Street and shortly you will be in Point Defiance Park, one of the largest urban parks in the United States, where a zoo, hiking trails, gardens, and beaches are bounded by the cold, clear waters of Puget Sound. Where Pearl Street dead-ends at the sound is a Washington State ferry terminal that provides passage to drivers, bikers, and pedestrians across the almost two-mile Dalco Strait to Vashon Island, about a fifteen-minute crossing.

Vashon Island is a small island in Puget Sound, home to over ten thousand people, a mix of farmers, environmentalists, hippies, artists, writers, and professionals who commute by ferry to Seattle.[2] Except for private boats, the ferry is the only way to get to the Seattle or Tacoma mainland because there are no bridges to the island. This has allowed Vashon to retain a rural, small-town feel despite its proximity to two of Washington State's largest cities. A ferry at the north end of the island connects Vashon to Seattle in a twenty-minute crossing. Jurisdictionally, Vashon is part of King County and is served by the Seattle–King County Health Department.

Vashon's natural resources were what first attracted Europeans to the island, which developed and grew with logging and berry farming in the nineteenth and early twentieth century. After World War II, as logging and farming declined and ferry service increased, the island increasingly became home to Seattle commuters. In the 1970s Vashon Island attracted more countercultural residents, artists, and hippies who wanted to get back to the land, grow their own food, and raise children in a safe and healthy environment.[3] The 1990s' technology and housing boom also brought change to Vashon. Today, multimillion-dollar Puget Sound and Olympic Mountain–view homes line some of the island's shoreline and ridges. Nevertheless, despite the profound changes that have occurred on Vashon Island over the twentieth century, the small island's rural feel and distinctive Northwest beauty even today are striking.

In the early 1970s, as Dr. Sam Milham was testing children in Tacoma and Ruston for arsenic exposure, Dr. Carl Johnson of the Environmental Health Division of the Seattle–King County Health Department was making inquiries at the local and state level as to what was known about the environmental accumulation of arsenic and lead on Vashon Island. Because the smelter's pollution was carried northeastward some nine months of the year due to prevailing winds, Dr. Johnson was particularly concerned with its impact on Vashon Island, and even at this early date researchers and health officials thought that Vashon might have an even worse environmental contamination problem than Tacoma.[4]

In the summer and fall of 1972 and the winter of 1973, PSAPCA and the Seattle–King County Health Department conducted environmental monitoring and human exposure studies on Vashon that documented higher than "normal" levels of arsenic in dairy cows, soil, and children. Some Vashon children also had elevated blood lead levels.[5]

The Vashon Island investigations added another dimension to local and state environmental and health agency attempts to delineate the geographic reach of the smelter's environmental contamination. By early 1973 the scientific findings painted a grim picture. Soils in Ruston, North Tacoma, and Vashon Island were found to be contaminated up to six to eight miles from the stack in the direction of prevailing winds; children and adults in Ruston had elevated urinary arsenic concentrations; Vashon children had evidence of both arsenic and lead absorption; urinary arsenic concentrations were highest in preschoolers; and health officials thought that the most important exposure was ambient air. Health and environmental officials also feared that food grown in Ruston and on Vashon Island might be contaminated with arsenic and lead.[6]

By the mid-1970s it was known to regulators that copper ore smelted at Tacoma contained the highest arsenic content of any U.S. smelter, and that the community was being exposed to heavy metals both from the smokestack and from fugitive sources.[7] The population living within the influence of the smelter's pollutants was among the largest of any copper smelter in the United States, and there were residences within several hundred yards of the stack, yet the Tacoma smelter consistently ranked low on the EPA's priorities.

The course that EPA headquarters should take was obvious to local public health officials, regulators, and advocates. Under the Clean Air Act, the EPA had the authority to regulate the Tacoma smelter's arsenic and other heavy metals emissions. To regulate arsenic, the EPA would first have to declare it a hazardous air pollutant and then set a standard aimed at protecting health within an "ample margin of safety." But the EPA's process for doing this was agonizingly slow. Between 1970 and 1990 the EPA set standards for just seven hazardous air pollutants.[8]

From the mid-1970s through the mid-1980s, Vashon Islanders began to play a significant role in calls to control the smelter's arsenic emissions. There was a strong environmentalist ethic on the island, and islanders saw themselves as recipients of much of the plant's pollution but with few of the benefits. Locals who increasingly called for federal regulation of arsenic in ambient air began a tussle with federal regulators that lasted well into the next decade. There were even divisions within the EPA, with the regional office in Seattle largely sharing local concerns about the problem, trying to secure funding for health studies, and pushing headquarters to set an arsenic standard.[9] Headquarters, however, repeatedly rebuffed the region and local regulators.[10] It questioned the strength of the evidence for arsenic's carcinogenicity and thought that more evidence of its effects on community health was needed to move forward with regulating it as a hazardous air pollutant.[11] However, it was inexplicably reluctant to fund such studies in Tacoma, the city with the highest air concentrations of arsenic in the country. This left local health officials and regulators in a very difficult position, without the political, scientific, technical, legal, and financial backing they needed from the federal government to protect the community from the smelter's heavy metal emissions. Paradoxically, local regulators were facing federal intransigence even though the EPA had been created to solve this very problem at the local and state level. And in the absence of governmental action, concerned community members in Ruston and Tacoma and on Vashon Island reverted once again to using the courts to try to gain redress from chronic pollution, with limited success.

The EPA's lack of alarm about arsenic in the mid-1970s contrasted sharply with the perspective of the National Institute of Occupational Safety and Health (NIOSH), the research arm of the Occupational Safety and Health Administration (OSHA). NIOSH had no doubt about the hazardous nature of arsenic, having determined in 1975 that it was a carcinogen and that worker exposure to arsenic in air should be as low as possible. By 1975 a number of U.S. worker studies had been added to the literature. Six studies of copper smelter workers and two studies of pesticide production workers had reported increased respiratory cancer risk attributable to arsenic. NIOSH issued a strongly worded statement in

1975: "Recent reports undeniably associate occupational exposure to inorganic arsenic with increased cancer mortality."[12]

In addition, two community studies suggested that arsenical air pollution might increase respiratory cancer rates in populations living near smelters.[13] In early 1975 OSHA proposed a 4 μg/m³ standard for arsenic in workplace air.[14]

At EPA headquarters, however, key officials were skeptical of NIOSH and OSHA's conclusions and met with OSHA in November of 1974 and "voiced our concern over the tenuous nature of data that links arsenic concentrations to adverse effects."[15] The EPA's view was that the epidemiologic studies of workers did not conclusively show that arsenic, not some other substance, was responsible for increased cancer incidence in workers, nor did the studies relate specific exposure levels to adverse effects. The EPA also thought that NIOSH had given too little weight to animal studies that did not "support the proposition that arsenic is a carcinogen."[16] The EPA's arguments were similar to those that ASARCO had been using to argue against a tighter workplace arsenic standard.[17]

But by the mid-1970s there was even local evidence that Tacoma smelter workers had an increased risk of lung cancer due to arsenic exposure. After being rebuffed for years in his quest for Tacoma smelter workers' medical records, Sam Milham examined Washington State death records by occupation and identified a twofold increased risk of respiratory cancer mortality among Tacoma smelter workers.[18] ASARCO, however, still did not admit to a link between arsenic exposure and respiratory cancer mortality. During this time workers at the plant produced a newsletter called *The Smelter Worker*, which provided a forum for discussion of health and safety concerns among other things. When Milham's findings were reported in the newsletter, it occasioned a letter from the smelter manager in which he noted ASARCO's long role in investigating adverse health effects of arsenic on Tacoma workers. He wrote, "So far we have found no evidence to establish such a connection" between airborne arsenic exposure and "respiratory cancer, heart ailments, and other disabilities." The letter also mentioned state toxicologist Ted Loomis's position that "he could not say for sure whether arsenic was or was not a weak cancer-causing agent because 'no chemical can be said to be absolutely safe in regard to carcinogenesis,' but that he was certain that arsenic was not a strong cancer-causing agent."[19] There is no doubt that some smelter workers were concerned about the health effects of arsenic exposure. They were also acutely aware of the tenuousness of their employment at the smelter, which was reinforced throughout the 1970s by threats to close the smelter if it was forced to meet stringent environmental standards.[20]

Ambient air concentrations of arsenic measured in 1973 in residential neighborhoods near the Tacoma smelter (4.2 μg/m³ three-month average and a maximum of 15.9 μg/m³ twenty-four-hour average) exceeded OSHA's proposed workplace standard.[21] This was problematic for the EPA, since environmental exposure standards are typically much more restrictive than occupational standards to account for the special vulnerabilities of community residents such as infants, children, and pregnant women to environmental toxics. People living near industrial facilities may be continuously exposed to toxics, in contrast to workers exposed during a typical eight- to ten-hour shift.[22] That OSHA had proposed a workplace standard that was at times exceeded in the community was a strong argument for taking decisive action to significantly lower the community's exposure to arsenic.

But before regulating arsenic in air, it seems that the EPA wanted definitive evidence of increased rates of cancer attributable to arsenic from community exposure—something that community epidemiologic studies rarely yield—rather than taking a precautionary approach and extrapolating from occupational studies.

At the time the EPA was supporting several studies of arsenic as a community air pollutant, but none included a focus on Ruston or Tacoma.[23] Without backup from the EPA, PSAPCA was in a difficult position to enforce regulations it had promulgated in 1973 to limit visible emissions of arsenic-containing particulates from the smelter's stack. And by 1975 the company was about to ask PSAPCA for more time to comply with key requirements.[24] PSAPCA felt that, as a local agency, it had neither the scientific expertise nor the political authority to set an ambient standard for arsenic. For this it looked to the EPA, the agency with the technical capabilities to scientifically defend such a standard from certain legal challenge.

The EPA, however, was awaiting a National Academy of Sciences (NAS) report on arsenic as an environmental pollutant, which it felt was essential to the environmental-standard-setting process. Requested in 1973, only a preliminary internal draft was available in early 1975, and the final report would not be published until 1977.[25]

Even without a federal standard for arsenic in air, PSAPCA's emission control requirements and enforcement did decrease arsenic emissions from the smelter's stack over the decade of the 1970s; however, these reductions in stack emissions were not eliminating children's exposure to arsenic in Ruston. Between 1972 and 1976 Sam Milham continued to monitor arsenic concentrations in Ruston children's urine, sampling children on at least fourteen separate occasions to determine whether concentrations were declining as arsenic emissions from the smelter's tall stack fell by almost 75 percent in the early to mid-1970s.

The emissions reductions occurred largely because of sustained pressure from environmental and public health groups as well as the responsiveness of local regulators. By the mid-1970s there was a fair amount of organized opposition to the smelter by groups concerned about its health and environmental impacts, nearly all of it originating outside of Ruston, in Tacoma, Vashon Island, and Seattle. This made for a very different political environment than in Bunker Hill. In Tacoma, the Washington chapter of the American Lung Association played a lead role throughout the 1970s until the plant closed. Also involved in the early 1970s were Clean Air for Washington, the Washington Environmental Council, and the Tuberculosis and Respiratory Diseases Association.[26]

Grassroots citizens groups from Tacoma and Vashon Island were also involved, such as GASP (Group against Smog Pollution), APE (Americans Protecting the Environment), IRATE (Island Residents against Toxic Emissions), the Sierra Club, and Friends of the Earth. These groups had a broader focus than simply controlling pollution from the smelter, and they situated their efforts to bring the smelter into compliance within the framework of a broader environmental and public health agenda.

The advocacy and oversight of public health and environmental advocates made it clear that the actions of regulatory agencies were being closely monitored. PSAPCA head Arthur Dammkoehler believes that his agency's regulatory accomplishments would not have been possible without citizen advocacy.[27] Additionally, legal challenges mounted by citizen groups established some important precedents for environmental law in the state.

Despite the reductions in arsenic emissions that occurred in the early 1970s, approximately 101 tons of arsenic per year were still being released in 1976 from Tacoma's tall stack. Low-level, or fugitive emissions—those that did not go through the stack—which were likely the cause of Ruston children's elevated urinary arsenic concentrations—still contributed an estimated 613 tons per year of arsenic to the Ruston environment in 1976.[28]

This was most likely why Sam Milham was still finding high average arsenic concentrations in Ruston children's urine in 1976, which led him to conclude that the controls installed at the smelter were not having the intended effects. In a letter to PSAPCA in October of 1976, Milham wrote: "For comparable groups, arsenic levels are about the same now as they were four years ago. If the extensive emission control activities undertaken by the Smelter have had an impact on overall community arsenic exposure, I cannot demonstrate it in this urinary arsenic data . . . I conclude that Ruston children are still being exposed to very high levels of arsenic in their environment."[29]

After 1976 the state's urine testing program appears to have stopped for several years. However, ASARCO began regularly monitoring children's urine by 1975, and for a time only ASARCO conducted such monitoring.[30] Discrepant results were reported at least once when both the State Health Department and ASARCO were testing.[31]

When ASARCO conducted urinary arsenic testing, it communicated directly with parents regarding their children's levels, and provided its own interpretation of the data. For example, after sampling in the fall of 1977, Curtis Dungey, the smelter's environmental specialist, wrote to a Tacoma parent whose child had an arsenic concentration of 184 μg/L in his urine, more than ten times what Milham considered normal, downplaying the child's exposure: "Small amounts of arsenic, such as this, show up quite rapidly in the urine when arsenic is ingested or inhaled. It does not mean that your child is gradually accumulating an abnormal amount of arsenic in the body. This is a normal function of the body—to eliminate a substance it doesn't need. A second sample, given on another day, may show your child's value to be below this average."[32]

In another communication with Ruston parents, Dungey claimed that Ruston children's urinary arsenic excretion was occurring at "very low levels . . . and do not mean a health risk is involved."[33] Ruston parents, many of whom were employed at the smelter and knew that their jobs were potentially at risk, were probably more likely to be swayed by these claims. ASARCO's communications with parents probably contributed to some of the ambivalence and confusion in the community regarding arsenic exposure, health risks, and the need for further regulation of emissions.

On the other hand, Vashon Island parents, whose livelihoods were not tied to the smelter, were not ambivalent about the smelter's pollution. They not only wanted strict controls on emissions, they wanted to know what heavy metal exposure might be doing to their children, and they demanded further testing. In the spring of 1976 ASARCO collaborated with the State Department of Health to test children's arsenic concentrations on Vashon Island. In late June the health department reported that there "was no abnormal urinary excretion of arsenic at the time the samples were taken."[34] However, the study found that urinary arsenic concentrations averaged 30 μg/L for five- to seven-year-olds and 36 μg/L among ten-year-olds, still over twice what Milham considered "normal" and about five to six times the average concentration found in unexposed communities that served as controls in a national smelter study.[35] Still, ASARCO issued a press release reporting on the results with the headline: "Tacoma Smelter Arsenic Emissions Pose No Danger to Maury and Vashon Island Residents."[36]

Although ASARCO tried to publicly downplay possible health consequences of children's urinary arsenic concentrations, even within the company there was apparently concern that some children were being highly exposed, enough concern for ASARCO to conduct environmental testing of some local children's homes and yards. In one instance the Drummond family's children, ages six, seven, nine, and ten, were found to have average urinary arsenic concentrations of 228, 202, 118, and 103 μg/L, respectively, in a week-long test conducted by the State Health Department in the summer of 1976. On one of the days, the six-year-old's urinary arsenic concentration was 800 μg/L, and the child had not eaten seafood in the two days prior.[37]

ASARCO wanted to determine "if anything around the home may contribute to the elevated urinary arsenic excretion of the Drummond children."[38] After analyzing air, house dust, building materials, and soil samples, ASARCO's chief chemist wrote to Dungey, the smelter's environmental specialist, "Possibly the reason for the children's higher level is the fact that they are playing in soils averaging 733 parts per million [arsenic] and not properly washing before eating. Are there fruit trees in the yard?"[39] Internally, Dungey described how the Drummond children were likely being exposed to the manager of the Department of Environmental Sciences:

> Playing outside in dirt and dust, or even inside on the floor may be the source of such arsenic absorption. Small amounts of arsenic from household dust may be inhaled when playing on the floor. This contact with arsenic dust both inside and out is possible because arsenic concentration found in the vacuum dust samples were of the same order of magnitude as those in soil samples. This follows, because most likely the dust found inside is the same as that from outside, whether it be brought in on shoes, or through cracks in windows, doors, etc. The arsenic could also be ingested from residue on unwashed hands after coming into contact with the dirt or dust. This, of course, would depend much on the personal hygiene habits of the children.[40]

In an environment contaminated with heavy metals, the youngest children are likely to have the highest exposures. They are at higher risk of both inhalation exposure (they breathe more air per pound of body weight) and ingest more heavy metals through hand-to-mouth behavior, which is most evident in children between six months and three years of age.

But during the 1970s and early 1980s, only limited urinary arsenic monitoring occurred among preschoolers, and health effects in this particularly vulnerable group were never comprehensively investigated. Of the almost two thousand urinary arsenic samples taken by ASARCO and the State Health Department in multiple years of testing, 80.8 percent of the samples were taken from six- to twelve-year-olds and only 9.6 percent from preschoolers.[41]

Although the State Health Department, and Sam Milham in particular, used the results from its urinary arsenic monitoring to argue for stronger controls on the smelter's arsenic emissions, the results did not provoke decisive action from other regulators or policymakers. In fact, after initial success in reducing arsenic emissions from the stack, local regulatory efforts were floundering by early 1976. PSAPCA's board had just granted ASARCO a reprieve from complying with local pollution control regulations for the next five years. Part of what appeared to sway the board was a study done for ASARCO by Philip Enterline, a biostatistician from the University of Pittsburgh. The study was a reanalysis of ASARCO's previous studies of cancer risk among Tacoma smelter workers. In 1975, for the first time, ASARCO would admit that its workers had an increased risk of lung cancer death that could be related to arsenic exposure.[42] The reanalysis showed that Tacoma smelter workers had a two- to eightfold increased risk of dying of respiratory cancer, consistent with the NCI's 1969 study of Anaconda smelter workers.[43]

Despite the increased cancer risk apparent in workers, when Enterline reported the results at a PSAPCA hearing, he emphasized the concept of a threshold for arsenic exposure among workers, a level below which arsenic would not cause lung cancer. The concept of a "threshold"—the idea that exposure to a carcinogen under a certain concentration will not cause cancer—is contested by environmental and public health advocates. Industry and its advocates are often the strongest proponents of the threshold argument, and environmentalists and public health advocates typically challenge the idea that any exposure to a carcinogen, no matter how small, is safe.[44]

At the PSAPCA hearing, Enterline testified that workers with urinary arsenic concentrations below 200 µg/L and exposed for less than twenty-five years did not appear to have an increased risk of respiratory cancer. An ASARCO lawyer questioned Enterline on the connection between his findings of an apparent threshold and what that meant for children living in the community, since children's urinary arsenic concentrations were mostly, but not all, below 200 µg/L. ASARCO's lawyer said: "Now, you've heard Mr. Varner (ASARCO employee) testify regarding the average urinary arsenic from children outside the Ruston School, and he testified that that average or mean was 65 micrograms per liter of urine; and I'm assuming then that would be well below what you would consider a threshold for 25 years exposure. Is that correct?" But Enterline, testifying for ASARCO, appeared hesitant to apply his findings to children in the community. He stated, "Well, I'm talking about a group of men who are 65 and over, and if that can be extrapolated to children, that is correct. That's quite an extrapolation."[45]

By the time the study was published in 1977, ASARCO scientists were less certain that 200 µg/L represented a threshold for arsenic exposure and cancer in workers and discussed this among themselves.[46] In the published paper, Pinto and his co-authors argued that the results indicated a threshold but that further work on establishing the threshold was continuing.[47]

Regardless, the threshold testimony and concerted pressure from ASARCO and the smelter's defenders as well as dismal economic conditions in Tacoma at the time all appear to have swayed PSAPCA's board. In early 1976 it voted to grant ASARCO five more years to comply with sulfur dioxide regulations and two to five more years to comply with local arsenic regulations.[48]

Throughout the 1970s regulators focused almost exclusively on adult cancer as the relevant health risk from smelter pollution in Ruston. There was little public discussion of health outcomes that might be more immediately relevant to children. In 1977, when University of Washington professor Dr. Ronald DiGiacomo proposed an intensive health study of area children that would look at the effects of arsenic on their nervous systems and their neuropsychological and cognitive functioning, ASARCO's environmental science director urged Tacoma smelter managers not to cooperate. Calling DiGiacomo's study "a waste of time and money," he stated, "If any active cooperation is requested . . . we should oppose the study tactfully at every opportunity. There is no lead problem. There is no prospect of uncovering an arsenic problem either because of the low dosage levels. Such levels have been exceeded in a number of locations around the world without detectable effects."[49]

The study was never conducted. Local and state officials were also unable to secure federal funding for a large-scale community health study, though they tried.[50] Tacoma was also not included in a national smelter study initiated by the EPA after the lead poisoning epidemic was discovered at Bunker Hill.[51] The decade of the 1970s would close without a federally funded or other large-scale study of community health effects from arsenic emissions around the Tacoma smelter, representing a missed opportunity to learn more about children's exposure to environmental arsenic.

It is interesting to note that although arsenic was known to be a carcinogen, in the 1970s and early 1980s arsenic exposure in children did not arouse the passion of the public health community as strongly as lead did. The lack of a scientifically established level of arsenic in urine that would indicate harm was part of the reason. As was pointed out by an occupational health scientist in 1975, the lack of scientific and public health attention given to children's arsenic exposure was likely a mistake. He noted that lead "appears to have captured almost exclusively the attention of some investigators studying the toxic effects of emissions from smelters on nearby communities," and argued that this was a critical oversight since arsenic emissions "appear to pose . . . a more severe environmental health problem—a carcinogenic one."[52]

The uncertainties about the effects of airborne arsenic on children's health, ASARCO's insistence that community health was not being harmed, and local officials and regulators who did not want to be blamed for job losses if the smelter were to close created the conditions that allowed regulators to prioritize ASARCO's economic interests over those of community health. The case was in a state of regulatory limbo at the local level; because of this, the only pathway to reducing the smelter's arsenic emissions in the late 1970s would have been a strong stance by the EPA.

It was 1977 when the NAS finally published its report on environmental arsenic. Not surprisingly, it concluded—as NIOSH had two years earlier, that "there is strong epidemiologic evidence that inorganic arsenic is a skin and lung carcinogen in man."[53] But even though the NAS thought that the evidence of arsenic's carcinogenicity was "strong," the EPA still considered the evidence inconclusive.

The director of the EPA's Air Quality Planning and Standards wrote: "Their [NAS] report fell short of an unequivocal statement regarding the effects of arsenic," and the EPA was still waiting for ongoing community health studies to make a decision on listing arsenic as a hazardous air pollutant. A study at a Baltimore chemical factory that used arsenic was extended for several years, a study of six smelting communities (but not Tacoma) was slated to begin in early 1977, and efforts were under way to characterize emissions and air concentrations of arsenic across the United States. The EPA estimated that a decision on regulating arsenic in air could "be made in mid-1978, when the results from the community health research projects will become available."[54]

But the EPA's own Carcinogen Assessment Group (CAG) read the NAS report differently, and on the basis of its conclusions, the CAG determined that the EPA should move forward with regulating arsenic. In April 1977 Dr. Roy Albert wrote to EPA leadership: "There is strong epidemiological evidence that inorganic arsenic is a skin and lung carcinogen in man.... While arsenic has not yet been shown to be carcinogenic in animals the evidence in humans is sufficient to warrant its being regarded as a carcinogen for regulatory purposes."[55]

Frustration with the EPA's slow pace on regulating air toxics even led Congress to specifically direct the agency, in the 1977 Clean Air Act amendments, to decide within a year whether arsenic would be listed as a hazardous air pollutant.[56] When the EPA missed Congress's 1978 deadline to decide on listing arsenic, the Environmental Defense Fund filed suit against the agency asking the court to order the agency to declare arsenic a hazardous air pollutant or detail why it would not.[57] It would take the EPA until June of 1980 to make the declaration. Almost a decade had passed since the regional EPA office, local regulators, and environmental groups had first asked for EPA headquarters' help on regulating arsenic. Even after taking the initial step of listing arsenic, a federal standard for smelters would take six more years and would come too late to help those exposed to the Tacoma smelter's arsenic emissions.

Two-time EPA administrator William Ruckelshaus once described the advent of the EPA and federal environmental regulation as akin to having a "gorilla in the closet."[58] By that he meant that states and localities now had powerful backup if they could not or would not regulate industries within their borders. But in Tacoma the EPA played no such role and by the 1980s prospects for protecting people in the Puget Sound region from smelter pollution were becoming more remote. The election of Ronald Reagan in 1980, who promised "regulatory relief" from "burdensome regulation," made a strong federal regulatory role in Tacoma even more implausible. Environmental and occupational health and safety regulations put in place in the 1970s were prime targets.[59] During much of the Reagan administration the EPA was mismanaged and leading officials were involved in high-profile scandals. The agency's administrator, Anne Gorsuch, eventually had to resign.[60] By the time she was replaced by William Ruckelshaus, serving his second term as EPA administrator, the agency was said to be on the "brink of disaster."[61] The disarray at the EPA in the 1980s worked in ASARCO's favor in Tacoma by extending agency timelines for regulating arsenic and giving the company a strong voice in the rule-making process.[62]

Meanwhile, in Ruston and Tacoma and on Vashon Island, residents lived with much uncertainty and ongoing pollution from the smelter. Since no comprehensive health studies had been conducted, questions and concerns about possible health effects could not be answered, contradicted, or validated with factual information. For those who were inclined to believe there was a problem, the lack of information created suspicion, which led to divisions in the community.

Concerned neighbors made complaints to the EPA and other agencies, such as one in 1981 after alleged dumping by ASARCO of material thought to be toxic near a Ruston residential property. The complainant's problems were summarized in the following way: "They have had all sorts of respiratory problems, they can't even breathe properly . . . her son developed bruise-like marks all over his body, one woman developed heart, liver and kidney problems, [her] infant grandson has been in the hospital at least 2 or 3 times with violent stomach cramps, diarrhea, and an inability to eat or drink, achy bones and joints have occurred, and people have reported holes in their noses."[63]

The complainant asked if smelter workers are wearing "gas masks, wouldn't people in the area need them too?"[64] This was a valid question since OSHA's final arsenic standard required that workers wear respirators when air concentrations of arsenic exceeded 10 $\mu g/m^3$—a concentration that was exceeded at times in residential areas of Ruston.[65] The lack of any community standard for arsenic exposure meant that smelter workers could be protected at work but exposed at home, and their families, if they lived nearby, could be continually exposed to arsenic in air twenty-four hours a day.

Throughout the final debates over regulating the smelter's emissions, the question of whether the public's health was being harmed remained a point of contention. A few epidemiologic studies were conducted, mostly by local investigators in the late 1970s. The studies relied on the analysis of routinely collected data, such as cancer incidence and deaths, and found little evidence of elevated cancer rates in the community.[66] Sam Milham conducted a study of the mortality experience of men who had attended Ruston elementary much earlier in the century. Milham concluded, based on a relatively small number of deaths, that "it seems not likely that the elevated urinary arsenic levels reported in 1973 by

Milham and Strong among Ruston Elementary School children will be of any future health significance."[67]

The findings of the community epidemiologic studies were not particularly surprising, nor did many health officials consider them conclusive since they were beset with methodological limitations. The lack of statistically significant findings did not exonerate arsenic from causing community health problems; rather, the lack of conclusive findings were just as likely due to methodological problems such as lack of accurate exposure measurement, small sample sizes, and lack of statistical power. Health outcomes relevant to children were also not systematically investigated in well-designed studies with adequate statistical power.

Relying largely on the analysis of secondary data and on some small studies of a few health parameters in Ruston schoolchildren, the State Department of Health appeared to become more convinced in the early 1980s that the health of those who lived in the vicinity of the smelter was not being harmed. In a strange flip-flop, the EPA became much more concerned about the smelter's health risks. The EPA attributed its newfound concern to the accumulating science on arsenic that consistently linked exposure to it with cancer in different exposure contexts (medical, occupational, environmental). Open disagreements between the two agencies broke out, which contributed to even more confusion in the community.

Meanwhile evidence of the smelter's environmental impacts continued to mount. In late February of 1981 all of the herring eggs in a thirty-thousand-square-yard area in Quartermaster Harbor, off of Vashon Island, were killed three days after they were spawned. Washington State Department of Ecology investigators pointed to arsenic in harbor sediments resuspended by sixty-five-mile-per-hour winds during a storm as a potential cause. Juvenile herring in the harbor were already known to be ailing—suffering from "broken back syndrome"—a condition that can be caused by heavy metal exposure during early stages of development.[68]

In the mid-1970s ASARCO was to have stopped discharging its process water into Commencement Bay. Before that it was discharging over 8.5 million gallons per day contaminated with 2,500 pounds of toxic metals including arsenic, antimony, cadmium, copper, lead, and mercury.[69] But a 1984 inspection found that ASARCO was once again discharging contaminated water into the bay, this time from a pilot slag granulation operation. The state ecology inspector recommended a $15,000 fine since "the violation has been so blatant and has been continuous for over two years." Once notified, "ASARCO made no effort to immediately stop or temporarily reduce the discharge. ASARCO didn't even give an explanation of their past actions."[70]

The Puyallup Indian tribe had strong concerns about the smelter's impact on Puget Sound fisheries. Frustrated by the ongoing pollution of Puget Sound, the tribe threatened to sue on a few occasions. A lawyer for the tribe stated in 1981: "Any corporation that impacts community health and affects tribal survival with the seeming impunity of ASARCO is a detriment to the whole community."[71] In the context of a variance hearing in late 1981 at which the company was requesting more time to comply with air standards, a Puyallup tribal councilmember said that if the variance were granted, the tribe would take the matter to the federal courts to stop "the 'disgusting degradation' of the earth."[72]

Despite the EPA's foot-dragging, the 1980s saw growing concern over the smelter's environmental impacts. Slag, which ASARCO had formerly dumped into Puget Sound, was becoming more of a concern in the early 1980s. The State Department of Ecology was finding that arsenic could leach out of slag. Slag was used in residential landscaping in Ruston and Tacoma. It was also used in sandblasting, paving, bridge building, shoreline stabilization projects, and cement-making. It was dispersed around the region and perhaps around the country, but a full accounting does not appear to have occurred.

The smelter's contribution to acid rain, which was a growing concern nationally and globally in the 1980s, was being investigated in the early 1980s. In the early 1970s ASARCO estimated that the smelter was putting out twenty-two tons of sulfur dioxide—a building block for acid rain—per hour at full capacity.[73] Acid rain causes acidification of lakes and streams, declining fish stocks, damage to agriculture, and damage to forests. The high-mountain lakes of the Cascade Range were thought to be particularly vulnerable to acid precipitation.[74] There was no question that the smelter's emissions were a substantial contributor to the problem in the region. The EPA predicted that there would be a very large reduction in acid rain in the region when the smelter closed.[75]

Still, environmental agencies were not taking further action to reduce the smelter's sulfur dioxide emissions due to a complicated set of regulatory circumstances. Congress inserted provisions in the 1977 Clean Air Act to give the nonferrous smelting industry more time to comply with sulfur dioxide emission regulations, and the changes also allowed for the possibility that dispersion, rather than sulfur dioxide capture, might be grandfathered in at some smelters. While the EPA was writing rules to deal with these changes to the Clean Air Act, it would not enforce existing federal sulfur dioxide standards in Tacoma. Because of this the EPA entered into a settlement agreement with ASARCO that put off federal enforcement of sulfur dioxide standards in Tacoma for the foreseeable future.[76]

The EPA settlement agreement limited PSAPCA's ability to enforce its own more restrictive 90 percent standard, and in 1981 the local pollution control board granted the smelter another variance from its sulfur dioxide control requirement effectively until January of 1987.[77] This meant that residents would have to live with sulfur dioxide pollution from the smelter until the plant closed.

In the early 1980s research continued on the geographic extent of the smelter's heavy metal pollution. Jerry Bromenshenk, a researcher at the University of Montana, was investigating a novel biological monitoring strategy using honeybees to determine the extent of environmental contamination with toxic metals. Bees forage widely, and in the process of collecting pollen, bring contaminants back to the hive, such as arsenic and other heavy metals that have settled on flowers. Electrostatic charges, which help bees to gather pollen, are also thought to cause pollution particles to adhere to their bodies. Arsenic and other heavy metals can also accumulate inside bees' bodies.[78] Bromenshenk thought that heavy metal concentrations in bees in the Puget Sound region could be used to map the geographic extent of smelter pollution.

Where concentrations of metals in bees returned to background, environmental concentrations would likely be at background too, providing an indication of how far the smelter's pollution traveled from Ruston. Bromenshenk also examined brood health and survival as another indicator of environmental damage.[79] With funding from the EPA he enlisted beekeepers ranging from Tacoma to as far north as Whidbey Island (an island in Puget Sound about ninety miles to the north of Ruston) to help with data collection.

Even Bromenshenk appeared surprised by his findings. Arsenic concentrations in Tacoma bees were sixty times background and "as high as we have ever seen in any of our previous studies and [were] more widely distributed in terms of the extent of the area affected." Arsenic concentrations in bees were high over a wide area, including South Vashon Island (12.5 parts per million), the West End of Tacoma (11.1 parts per million), and south Seattle (5.5 parts per million), some twenty-five miles north of Ruston. Arsenic concentrations of 4 parts per million and higher are considered hazardous to bees.[80] More than 64 percent of the colonies studied in the region exhibited low brood viability. At the time Bromenshenk wrote that the highest arsenic concentrations he had ever measured were in bees near Commencement Bay.[81]

Another study to indicate the wide geographic spread of the smelter's heavy metal contamination was one of garden soils conducted by the Tacoma–Pierce County Health Department in 1983. Concentrations of arsenic and cadmium in gardens were elevated for up to five miles from the smelter. The study also concluded that there was a potential for adverse human health effects from the concentrations of arsenic and cadmium found. The bee and garden soils studies only amplified concern on Vashon Island, where many people farmed or gardened and fed their families with homegrown produce.[82]

Environmental regulation at the state and federal level was meant to prevent, mitigate, and solve pollution problems on behalf of the public. The creation of environmental laws governing air and water pollution resulted in part from the recognition that society at large has an interest in a clean, healthy environment and that pollution travels—a local source of pollution can easily cross jurisdictional boundaries. Prior to environmental regulation, individuals affected by pollution had little recourse—they could live with pollution, try to get compensation for damage, or challenge polluters in the courts (typically using nuisance laws), where they could easily be outspent by industries with deep pockets. In Tacoma in the early 1980s, local residents, frustrated with the slow pace of government regulatory action, once again turned to lawyers and eventually to the courts to try to get redress.

A July 1981 fire in the smelter's smokestack spread fallout that contained up to 14 percent arsenic and 0.5 percent lead throughout the community.[83] Yet it took the Tacoma–Pierce County Health Department three weeks to warn gardeners that their produce might be contaminated. After an article on the stack fire appeared in the newspaper, several hundred people called the health department with their concerns.[84]

A Ruston couple, the Wingards, filed suit against ASARCO after the stack fire. Their organic garden was described as "once prodigiously fruitful, producing nearly enough for their own food needs and more to sell at local markets." After the stack fire, many of their plants died. Jean Wingard said, "Our whole life has changed, I don't think it will ever be the same here."[85]

Tests found high levels of lead, arsenic, and cadmium in their soil. The Wingards tried to sell their house, but it languished on the market for twenty months with no offers. The Wingard's lawyer characterized ASARCO's response as pulling out "all the stops in fighting the case," concerned about what a victory by the Wingards could mean for its liability to other residents. The company "imported expert chemists and doctors from out-of-state to testify that any damage to the Wingard's property has been minimal at most."[86] The Wingards lost.

Another accident in January of 1982 released hundreds of pounds of arsenic-contaminated dust that spilled from a storage bin and covered a car and a section of roadway near the smelter. Twelve people driving on the road who were exposed to the dust were taken to the hospital—including a six-week-old baby.[87] The dust contained about 48 percent arsenic. One couple, whose car was covered in the dust, hired a lawyer to persuade ASARCO to replace their $2,200 car. Maintaining that the arsenical dust "would only be harmful if you choked to death on it," the company's lawyer proposed that the couple should continue to drive their car but use an air monitor.[88]

These incidents did not provoke any exceptional reaction from local, state, or federal regulators. In fact, the EPA was actually engaged in negotiations with ASARCO to not promulgate an arsenic standard.[89] The State Health Department was growing less concerned about human health threats from the smelter, and the local health department had few resources available to them and did not think that addressing smelter pollution was its role. For the community, there was no government agency to turn to that could or did provide tangible help.

Bill Tobin, a lawyer, had lived on Vashon Island since the early 1970s. He was alarmed by heavy metal contamination on the island from the smelter's emissions and was waiting for the right time to bring a lawsuit against the company. For many years he lacked the funding and scientific evidence to do so. But by the early 1980s a citizen's group lawsuit had led to the compilation of an environmental impact statement on the smelter's continuing operation. The statement provided a rich source of information for Tobin. He had already formulated the legal strategy he would pursue, "under . . . two common-law theories—nuisance, which is the interference with the reasonable use and enjoyment of land, and . . . trespass to your property." Tobin thought that if he could convince the state court that trespass could include wind-blown toxicants, he could establish ASARCO's liability for contamination throughout the region—which was potentially enormous.[90]

He presented the idea at a community meeting in 1983, and within a few months Mike and Marie Bradley, property owners on the south end of the island, signed on. The Bradleys wanted ASARCO to pay a small amount of money to replace soil in their garden. Tobin thought that because of the small amount of money involved, the case, which he brought in King County Superior court, would be resolved through mandatory arbitration. However, he greatly underestimated ASARCO's response. Tobin said filing the case against the company was like "hitting a hornet's nest."[91]

ASARCO was taking the case very seriously, most likely because of the liability that a Bradley victory would imply. If the Bradleys won, there would be nothing to stop people on Vashon or in Tacoma with contaminated soil on their properties from suing the company for toxic trespass.

In the end, Tobin and the Bradleys convinced the court that smelter emissions that blew over the Dalco Strait and settled on the Bradley's property constituted trespass. The court found that ASARCO had known "for decades" that toxic metals were being released from its smokestack and that they had the potential to contaminate property, thus ASARCO had the "intent to commit intentional trespass." Further, "the defendant's conduct in causing chemical substances to be deposited upon the plaintiffs' land fulfilled all of the requirements under the law of trespass."[92] But the court also ruled that to win compensation, the plaintiffs would have to show that ASARCO's toxic trespass had caused damage either to their land or their health.[93]

Tobin recalled, "This is where our things started falling apart for our case." Proving damages was much more difficult than proving that ASARCO's pollution had trespassed on the Bradley's land. The state of the science on environmental arsenic was part of the problem. There was no environmental standard for arsenic in soil at the time in Washington State or federally. The Seattle–King County Health Department had advised against gardening in contaminated soils, but it was apparently reluctant to participate in the lawsuit. With no documented health effects from exposure to contaminated soil and an inability to point to specific environmental damage, the case was dismissed. Despite the fact that the Bradleys were not compensated, Tobin takes pride in having established ASARCO's liability for toxic trespass. Tobin believes the case "broke the dam open as far as liability is concerned."[94]

Another resident tried unsuccessfully to sue ASARCO in April of 1986—apparently the first lawsuit that explicitly charged that smelter emissions had caused health damages. David Reed alleged that exposure to arsenic from the smelter caused him to suffer from cognitive problems and medical conditions, such as stomach problems and numbness in his hands. Reed lived within several blocks of the smelter and had done significant landscaping in his yard. A soil test found arsenic concentrations in Reed's yard up to 2,520 parts per million—extraordinarily high levels for a residential yard. He had been diagnosed with peripheral neuropathy, which can be caused by exposure to arsenic, and had once had a urinary arsenic test that showed a high concentration. ASARCO's experts testified that Reed's health problems were preexisting due to a back problem, and that his brain and nerve functions were normal.[95] Reed lost his case.

For its part, ASARCO continued to advocate in various venues for the relaxation of environmental laws that applied to smelters. Speaking at a conference sponsored by the Conference Board and the U.S. Bureau of Mines, ASARCO chairman Charles Barber told attendees that environmental regulation was threatening the smelting and refining industry in the United States. Barber argued: "If the U.S. is to retain its surviving smelters, some accommodation must be provided in existing environmental imperatives." He called for the Reagan administration to achieve a "better balance between the competing national goals of environmental protection and preservation of U.S. smelter capacity."[96]

But much of the industry's pushback was a way of buying time since many of the nation's oldest smelters were of limited usefulness and their days were numbered. By the 1980s the primary nonferrous smelting industry in the United States was going through upheaval. Many smelters had closed, and the industry and its supporters sought, in part, to blame the high costs of complying with environmental regulations. But the industry had been in a slow decline since World War II for a number of reasons, with foreign competition one of the most important factors.[97] For Tacoma, it was becoming less and less feasible to ship ore from mines abroad.

The Tacoma smelter's niche—high arsenic ore—had contributed to its longevity. Public health advocates had repeatedly suggested that to dramatically decrease arsenic emissions, ASARCO should be forced to stop smelting high arsenic ore. But high arsenic ore was still profitable for ASARCO, even in the early 1980s, in part due to the high gold content of the ore.[98] The company knew that its main source of ore from the Philippines would not last much longer since a smelter was being built there to process it. By early 1986 the Philippine ore shipments would cease. Meanwhile, the arsenic recovered from Tacoma's smoke stream was finding its way into consumer products and being dispersed in the environment in applications such as wood preservation, cotton herbicides, and cotton desiccants, contributing to new problems.[99]

The EPA also continued to move slowly on regulating arsenic in air. In June of 1980 the agency at last declared arsenic a hazardous air pollutant, noting a strong link between it and skin and lung cancer. Nonferrous smelters were thought to be responsible for an astounding 65 percent of arsenical air pollution in the United States, and the Tacoma smelter alone was responsible for fully one-quarter of the country's airborne arsenic emissions.[100]

After declaring arsenic a hazardous air pollutant, the EPA was to regulate it by developing a National Emissions Standard for Hazardous Air Pollutants (NESHAP). An initial draft was abandoned, over the strenuous objections of a federal advisory committee, on the grounds that a national standard was not needed since, after the closure of the Anaconda smelter in 1980, Tacoma was the only remaining high arsenic feedstock smelter in the United States. Rather than setting a standard, the agency proposed entering into an agreement with ASARCO to install pollution control upgrades at Tacoma.[101]

The approach angered local activists and PSAPCA who were looking to the EPA to finally regulate arsenic. Gene Lobe, the board chairman, in a frustrated letter to Administrator Gorsuch, wrote, "The tens of thousands of persons living within 16 kilometers of the Tacoma smelter deserve better treatment of this issue by EPA than EPA headquarters apparently proposes."[102]

The Puget Sound region, however, was not the only part of the country with a stake in the arsenic NESHAP. New York State was also concerned about arsenic because it was present in emissions that reached its borders from glass manufacturing plants in New Jersey. New York sued the EPA in 1982 for not issuing the arsenic NESHAP, and a U.S. District Court ordered EPA to promulgate a standard within 180 days. The new deadline was July 11, 1983.[103]

By this time, William Ruckelshaus was heading the EPA for a second time (he was the first EPA administrator). In between terms at the EPA, Ruckelshaus had lived in the Puget Sound region, working as a senior executive at Weyerhaeuser, then based in Tacoma.[104] He was likely well aware of at least some of the contentious history around the smelter.

The agency he took over in 1983 was beleaguered from mismanagement, staff losses, and budget cuts, and the deadline for issuing the arsenic NESHAP now belonged to him. A few days ahead of the deadline, in early July, the EPA announced that it was finally poised to propose a standard for airborne emissions of arsenic. The national press reported that the standard would be less stringent than some in the EPA and environmentalists expected.[105]

Rather than setting a standard based on protecting health as the statute required, the EPA proposed using the "best available technology" standard to control emissions.[106] The application of this standard meant only minor upgrades at Tacoma, which ASARCO had already promised local regulators it would implement. The EPA accepted ASARCO's contention that most sources of arsenic emissions in the plant were controlled at the time using best available technology.[107]

The proposed standard would allow Tacoma to continue operating regardless of the ambient air concentrations of arsenic in the community, even if it continued to exceed occupational standards. The NESHAP would only result in an estimated 20 percent reduction of air emissions of arsenic in the United States.[108]

Computer modeling done by the EPA predicted that significant cancer risks from arsenic in the air would remain for Tacoma residents.[109] Because of this, Ruckelshaus proposed that the EPA hold an extensive public comment period, including community meetings, to discuss whether Tacomans favored stronger controls, which would likely result in the smelter's closure, and "whether a regulation which could cause a closure of the plant is appropriate in this instance in order to reduce the risk of lung cancer in Tacoma."[110] Essentially, the EPA was asking local residents to choose between jobs or more stringent regulation that would better protect their health.

This troubling choice was also put to Tacomans at a time when the city was still suffering from high unemployment and other economic woes. In 1982 the country was in a recession with a 9 percent jobless rate, which rose to 10 percent by the end of the year amid ongoing factory closings and layoffs. In Washington State the unemployment rate rose to 13 percent, with blue-collar workers particularly hard hit. Boeing, a large area employer, had been cutting jobs since the early 1970s and in the early 1980s got rid of another eleven thousand jobs. Tacoma was undergoing a difficult transition away from manufacturing jobs to more employment in service industries and high-tech, which would last well into the 1990s.[111]

The *Tacoma News Tribune* wrote that Tacomans "will be making history" by helping the EPA to decide on acceptable risks. The *Tribune* asserted that the plant's risks were "theoretical" and reminded the community that "Asarco, like it or dislike it . . . has been a part of the community for many years. It is more than just an awesome smokestack and eerie glows in the night. It is people, almost six hundred of them who work there and produce commodities which are necessary and beneficial to the national well-being."[112]

Outside of Tacoma, however, the response was much more critical. The *Seattle Post-Intelligencer* implied that the EPA's approach amounted to "weighing human lives against dollars."[113] Under fire, Ruckelshaus explained his rationale for putting this choice to Tacomans in the *New York Times*: "For me to sit here in Washington and tell the people of Tacoma what is an acceptable risk would be at best arrogant and at worst inexcusable." The *Times*, unconvinced, said, "Mr. Ruckelshaus has it all upside down. What is inexcusable is for him to impose such an impossible choice on Tacomans."[114]

Unacknowledged was that Tacomans had been in dialogue with the company and the city regarding the smelter's pollution for over sixty years, and with the federal government for nearly thirty years. During those sixty years, concerns about the cost of regulation and the impact on area jobs had repeatedly stymied efforts to regulate the smelter.

In fact, it was the recognition that industry's concerns about the cost of pollution control had long paralyzed progress on cleaning up air pollution that led Congress in section 112 of the 1970 Clean Air Act to direct the EPA to set standards for hazardous air pollutants within an "ample margin of safety," without regard to the cost of doing so.

There was a chorus of industry opposition to section 112 throughout the 1970s and 1980s. The basic argument was that section 112 was an overreaction to irrational fears of health risks from pollution that could devastate U.S. industry. The EPA interpreted section 112 for carcinogens as allowing no emissions since no level of a carcinogen could be proven to be safe. The EPA warned that section 112 could lead to the shutdown of major industries.[115] The agency did little to enforce the law in the 1970s and 1980s. Two decades worth of regulatory effort on hazardous air pollutants were essentially lost.

Ruckelshaus and others argued for a cost–benefit approach to making decisions about controlling hazardous air pollutants and advocated for risk assessment coupled with assessing the costs and benefits of pollution control as cornerstones of EPA decision making.[116] Key proponents of cost–benefit decision making viewed the health-based approach to environmental decision making codified in parts of the Clean Air Act as irrational, technically infeasible, and potentially economically ruinous. They argued that since all of life entails risk, and since risk from pollution is also coupled with social benefits, regulators must be careful not to overstep and impose unrealistic costs on industry. Risk assessment, or the science of quantifying risk from environmental hazards, would provide the scientific basis for evaluating the risks society faced, and cost–benefit analysis would provide managers with the tools to craft sensible regulation.

After relatively limited involvement in Tacoma smelter issues under Gorsuch, the EPA became heavily involved under Ruckelshaus. Federal regulation of arsenic from the Tacoma smelter became a high-profile test case for implementing risk assessment and risk management principles coupled with cost–benefit analysis. While under court order to issue a regulation on arsenic, Ruckelshaus asked Tacomans to make a choice between jobs with accompanying arsenic emissions and stringent health-based emissions standards and, most likely, closure of the smelter. In tallying the costs of ASARCO's pollution, however, the EPA only considered excess lung cancer risk in the area. The agency did not try to determine how much the smelter's pollution had cost the region in terms of property contamination, respiratory disease, other cancers, and chronic disease from arsenic exposure as well as the costs to children from reduced life chances due to heavy metal exposure.

The case presented the opportunity for a dramatic conflict between those who demanded a clean environment and those who needed industrial jobs to support their families.[117] Tacoma was chosen despite the fact that Tacoma's fate, by this point, was basically sealed.

The call for community involvement in the arsenic standard, however, came at a time of heightened public concern—particularly among Vashon Islanders—about smelter emissions. The Vashon Island Community Council wrote of its frustration with ASARCO's ongoing pollution in a letter to Ruckelshaus: "The burden of proof falls on us to show that elevated levels of arsenic are unhealthy, essentially making us unwilling guinea pigs in a toxic tolerance program."[118]

When the EPA held public hearings to get input from Tacomans, the discussions revealed how little the EPA actually knew about potential health effects of arsenic exposure in the community. Residents wanted to know: Were pregnant women or their fetuses at risk? Were there health hazards for children exposed to arsenic in soil? Was eating vegetables grown in contaminated soil dangerous? But agency scientists could not answer these questions with any degree of specificity, and they knew even less about the health hazards posed by cadmium in the area.[119] In the absence of scientific answers, some put stock in the claims of Ruston residents who asserted, "I've lived here 61 years, I am 88 years old and still breathing."[120]

There was a confusing array of opinions on the health risks of the smelter's arsenic emissions. EPA scientists asserted that because arsenic was involved, there was certainly a health risk. Workers afraid of losing their jobs questioned scientific experts, who often contradicted each other. The State Department of Health and the EPA engaged in public and spirited disagreement over the health risks to the community. The regional EPA office in Seattle appeared much more concerned about the smelter's arsenic risk and impact on the community than did headquarters. ASARCO and its experts added another dimension to the disagreement by emphasizing the threshold argument as an explanation for why no pattern of increased cancer deaths was discernible in the community epidemiologic studies—air concentrations of arsenic simply weren't high enough to cause lung cancer or other health problems.[121] The debate took place publicly and was covered in the local media, which only added to the confusion.

Doug Sutherland, Tacoma's mayor, appeared on *Good Morning America* in August and called ASARCO a "good corporate neighbor," and said: "Until I've been able to be shown specifically to me that there is indeed deaths being created by the emissions out of ASARCO, I don't think it should be closed."[122] Philip Landrigan was one of the few high-profile health experts who took a strong public, precautionary position by stating, "To assume the existence of a cancer threshold for a community population would violate prudent public health practice."[123] Much was made in the press of the disagreement and uncertainty among experts, and it certainly affected the credibility of health officials, particularly EPA scientists.

Once again, public relations played a key role in ASARCO's response. The public relations firm Hill & Knowlton managed its communications strategy and detailed the plans in a confidential memo. Hill and Knowlton advised that "THE key issue is the potential health hazard from arsenic emissions at the plant—all other issues are minor by comparison." ASARCO was urged to "acquire the services of a third party 'expert consultant,' credentialed and fully capable of speaking to the issues of 'risk assessment' and 'health risks related to arsenic exposure.'" The expert would be "booked for as many news media interviews (print and electronic) as possible." Supporters in the business community would be asked to attend the hearings and testify on ASARCO's behalf. Hill & Knowlton also recommended holding "a pro-ASARCO rally on the part of employees, families, community and business leaders to offset or balance the antics of our detractors. *Of course, it is essential that there be no public perception of ASARCO having a direct hand in organizing such an activity*" (emphasis in original). Other activities were to prepare press kits, arrange meetings with newspaper editorial boards and reporters, and provide background materials to reporters.[124]

ASARCO's expert emphasized the threshold concept with respect to arsenic exposure and pointed to recent a study conducted by University of Michigan scientist Ian Higgins and funded by the Anaconda Company. The study found that relatively high levels of arsenic in air were necessary to cause cancer in workers. Since exposures in residential areas of Ruston and Tacoma were much lower, the implication was that the community had little to be concerned about.[125]

Eventually, after the hearings were long over, a State Department of Health study would conclude its data was suggestive of a small increased risk of lung cancer among women who lived close to the smelter for more than twenty years. The authors stated: "Such an excess risk may be expected since even during the period 1980 to 1985, community 24-hour ambient air arsenic exposure exceeded the 8-hr work place standard of 10 $\mu g/m^3$."[126]

By the time of the public hearings on the arsenic standard, a consensus among community groups in Tacoma and on Vashon Island was emerging around the need for tighter controls on smelter emissions, with state and local public health departments, environmental groups, the State Department of Ecology, PSAPCA, and Tacoma city government favoring the EPA's setting an ambient air standard for arsenic. At the public hearing, despite doubts about significant community health effects, Milham testified in favor of a twenty-four-hour air standard. Dr. Bud Nicola, head of the Tacoma Pierce County Health Department, made a distinction between risks that people choose (like smoking) and risks that are involuntary (like exposure to industrial pollution) and argued that involuntary risks are up to society to control. He called on the EPA to establish an environmental and physiological monitoring program and to establish "safe ambient air standards for the community."[127]

The consensus position was a rejection of the EPA's approach, which presented regulation as in conflict with jobs. The public, by and large, wanted both health and jobs. And the EPA's view on the smelter's risk, which only considered lung cancer, had little relevance to how people saw their own risk, and to the risks many were most concerned about—elevated urinary arsenic concentrations in children. Vashon Islanders thought a good "standard" would be that the company should be allowed to emit no carcinogens—an indication of how communities were redefining risk consistent with their own concerns and values. Many residents took a more holistic view of risk than the EPA's lung cancer models did, expressing concern about health effects for pregnant women, infants, and children, and about exposure to arsenic through soil and water as well as air.[128] Predictably, many workers and their families who saw family wage jobs disappearing supported more minimalist standards, hoping that such standards would allow the plant to keep operating.

For all of the effort, controversy, and strife, the Tacoma hearings did not result in the EPA regulating arsenic emissions from the Tacoma smelter. Facing a July 1984 deadline set by PSAPCA to agree to install either flue gas desulphurization or a new smelting furnace to control sulfur dioxide emissions, ASARCO announced in June of 1984 that it would close the smelter within the year.[129]

For residents of the Puget Sound region concerned about smelter pollution in the 1970s and early 1980s, increased regulatory authority at all levels of government did not bring quick answers or decisive improvement to the community. Until ASARCO closed in the mid-1980s, local children continued to excrete arsenic in the urine, and sulfur dioxide emissions did not approach the 90 percent control that PSAPCA ordered in the early 1970s. Community members who were concerned about pollution had little recourse. They could either move or continue to live with smelter pollution until the plant closed. In Tacoma, the promise of a cleaner environment as an outcome of local, state, and federal environmental regulation was largely not to be realized during the 1970s and well into the 1980s. Tacoma demonstrates the forces and circumstances that eroded this bold social vision.

7

Sacrificed

> "Dust on houses over a mile from the [Tacoma] smelter still have high levels of arsenic, and unfortunately, always will."
>
> –Kevin Rochlin, EPA Tacoma smelter site manager, to Ruston mayor Bruce Hopkins[1]

When summer finally comes and the blanket of clouds that seems to cover the Puget Sound region all winter and spring lifts, if you stand on a hill in Ruston and look out to the north and west, it is easy to forget that you are standing on a site where an environmental disaster slowly unfolded over a the space of a century. The blue Puget Sound sparkles in the sun, bright white sailboats dot Commencement Bay and Dalco Passage, the snow-covered Olympic Mountains jut out of the sky to the west, and the green rolling hills of Vashon Island are visible to the north. On the former smelter site, developers are working to transform it, building high-end retail and housing that will take advantage of spectacular water and mountain views.

It is hard to imagine a time when smoke billowed from the smelter fouling the air across the region. The gloomy days of the mid-1980s when ASARCO closed Tacoma for good, putting about five hundred people out of work, seem like a bad memory.

The Tacoma smelter closed right around the time that the EPA's regional economist predicted that it would. In a 1981 report he judged the smelter would last four to five more years. The external factors impacting Tacoma were affecting primary nonferrous smelters all over the West: competition from the developing world (cheaper labor and easier access to higher-grade ore), aging facilities, and domestic economic policies.[2] Although many at the time placed blame on environmental regulation, the EPA's economic analysis of the Anaconda and Tacoma closures found that factors such as changes in international copper markets and competition, rather than environmental regulation, were the driving forces perpetuating those closures.[3]

Regardless, environmental and occupational safety regulations were blamed in part and helped to pit workers against environmentalists, and affected communities against the EPA. In Tacoma much was made of the role of environmental regulation in the closure, and the local steelworkers' union even asked the EPA to review the factors that led to the closure and to determine the role that regulation played.[4]

According to the EPA's economist, the main problem for the Tacoma smelter was that it was losing its lifeblood—cheap ore, at that time high-arsenic Lepanto ore from the Philippines. The Filipinos were building their own smelter to produce copper and other metals domestically. Tacoma's tidewater location had been beneficial in the past, but it was too far away from reliable ore sources by the 1980s. Because of this and other factors, ASARCO was consolidating its mining and smelting operations in the Southwest by building a modern refinery in Amarillo, Texas, and upgrading its smelter in Hayden, Arizona. By the 1980s Tacoma was a relic of another world order.[5]

This analysis of Tacoma's closure led the EPA to conclude that ASARCO had known since the mid-1970s that Tacoma would not be viable in the long term and therefore never intended to make significant investments in pollution controls to keep the smelter operating. Therefore, "disagreements over the cost of compliance served as no more than bargaining levers in the protracted negotiations between the firm and the regulators." EPA saw no role for air pollution regulations in Tacoma's closure, instead concluding: "Its roots lie in the deterioration of American heavy industry and basic changes in our world trade position. EPA is not the appropriate institution to deal with those matters."[6]

Even retired ASARCO chairman Charles Barber did not dispute this analysis. He admitted that it was clear to him that Tacoma was "doomed" in 1975, and that the company delayed investment in air pollution control in an attempt to prolong operations. But he decried the "cynical tone" of the EPA's report and said he was "proud to be associated with a company that could invest so much for the next generations."[7]

In the end, the EPA told the steelworkers' union that it would not study the matter further or hold public hearings on Tacoma's closure.[8] The union asserted that it would seek changes in federal laws so that "companies in economic trouble will receive federal help, such as tax incentives, and EPA assistance in achieving compliance with pollution standards."[9] The conflict helped to further the atmosphere of divisiveness between workers and environmentalists and perpetuate the perception that environmental regulation was unacceptably costly.

In fact, the EPA's protracted rule-making and compliance deadlines were more than forgiving to the smelting industry even though delays and inaction put public health at risk in Tacoma, Bunker Hill, and El Paso. In Tacoma the delays meant that ASARCO never had to meet a federal standard for arsenic because an arsenic standard for copper smelters was not promulgated until after Tacoma closed.[10] At no time when the smelter was operating did the company have to meet a limit on the concentration of arsenic it could put into Tacoma's air. It also never met the local sulfur dioxide standard, and its record of complying with the lesser federal standard was inconsistent.[11] The reductions in arsenic and other heavy metal emissions that occurred over the 1970s and early 1980s were due to local regulatory efforts, which did not go far enough to prevent children from ongoing, measurable arsenic exposure.

It was March of 1985 when copper smelting in Tacoma ceased for good. The arsenic plant stayed open to process the remaining arsenic and closed completely in 1986. Until the smelter and arsenic plant closed, arsenic concentrations in air in Ruston on some days exceeded the occupational standard.[12]

If the 1980s marked the close of the copper century, as historian Michael Malone has written, it also saw the continuing decline of primary lead smelting in the United States. In 1978 there were six primary lead smelters left, and by 1990 only the Herculaneum and Glover smelters in Missouri and the East Helena smelter in Montana remained.[13]

Bunker Hill was part of this trend, closing in 1981 amid an acrimonious debate over worker and environmental health standards and allegations of federal government overregulation and overreach. But, like Tacoma, Bunker Hill's closure was driven by many factors including aging facilities, which made competing internationally and meeting environmental and occupational health standards nearly impossible; decreases in silver and lead prices; and its location, far from a port.[14] The community, long dependent on mining and smelting, was turned upside down.

But even the Bunker Hill smelter, the cause of the United States' worst community lead poisoning disaster from an industrial source, was allowed to operate until closure without meeting federal air quality standards for lead.[15] Although the EPA issued an ambient lead standard of 1.5 µg/m^3 in 1978, the agency did not expect compliance until 1982. When announcing the new lead standard the EPA signaled its willingness to revaluate its regulatory approach if regulation would cause "major disruption" to the smelting industry.[16] Sulfur dioxide pollution also continued until Bunker Hill closed since the EPA entered into a settlement agreement in 1979 with the company that allowed the continued use of tall stacks and dispersion.[17]

The El Paso smelter remained in operation longer than Tacoma or Bunker Hill, probably owing to its location in the Southwest, closer to reliable sources of ore. El Paso air continued to register exceedances of the federal lead standard until the lead smelter closed in 1985.[18] Although the El Paso copper smelter continued to operate, ASARCO did not have to worry about meeting a limit on arsenic in air since the federal arsenic standard promulgated for copper smelters in 1986 did not require this. When El Paso closed in 1999 toxic metal emissions had been substantially reduced over early 1970 levels, but they were not completely abated. In 1994 arsenic concentrations were measured at an El Paso elementary school, two miles from the smelter, ranging from 0.034 to 0.388 µg/m^3, and the company later agreed to reduce arsenic concentrations tenfold.[19] Recently ASARCO applied for a permit to reopen the smelter and received significant push-back from former workers, community members, and elected officials. The smelter is now closed for good and is being dismantled and demolished.

Although some environmental and public health activists had high hopes that smelter closures would solve chronic environmental health problems, the closures shined a spotlight on the widespread environmental damage left behind and lingering human health threats, and they exacerbated anxieties and social strife about the future of smelting communities. Towns like Ruston, long dependent on one major employer, reeled from job losses and the loss of tax revenue. The contentious debates over health threats from smelter emissions shifted to arguments over the health threats posed by the accumulation of heavy metals in yards, gardens, parks, and other public spaces.

New questions arose, over which industry, government scientists, and communities would spar: How much environmental cleanup, if any, should be done? What were the health risks of living in an area contaminated by a century's worth of toxic metal pollution? How do communities live with and cope with chronic pollution? Can such persistent and widespread pollution ever be cleaned up?

In Tacoma, residents were coming to the painful realization in the early 1980s that their city rivaled many others west of the Mississippi as far as the extent and complexity of industrial pollution in Commencement Bay and the tideflats, which would be ranked in the top ten of the country's highest priority hazardous waste sites in October 1981 under the new federal Superfund program.[20] A complex mix of toxic chemicals was polluting the bay, with arsenic and heavy metal contamination from the smelter contributing significantly. Bay sediments near the smelter were found to be nearly devoid of plant or animal life.[21] In Ruston and Tacoma, the smelter site's inclusion on the Superfund list marked the beginning of another long and controversial chapter of this story, which continues today.

Just as many western smelting companies balked at controlling their pollution while operating, some also resisted paying for comprehensive Superfund cleanup and used an array of strategies to limit their costs. Gulf Resources transferred corporate assets and declared bankruptcy in 1994, and the federal government was only able to secure $18 million from Gulf for Silver Valley cleanup from the settlement.[22] When Idahoan Robie Russell was appointed to head EPA's Region 10 in the mid-1980s, he became the focus of an EPA inspector general's investigation for stalling formal enforcement actions at the Bunker Hill site. In the meantime, the Bunker Limited Partnership (BLP), which by then owned the smelter site, was able to transfer assets, an action that the inspector general feared would complicate the agency's ability to recover Superfund costs.[23] BLP, too, eventually declared bankruptcy.[24]

In Tacoma, after the EPA rejected an initial offer from ASARCO to spend only $1.1 million on remedial actions, the company began an extensive public relations campaign to influence Puget Sound residents, elected officials, community leaders, and the EPA to support its more modest cleanup proposals, apparently spending more money on public relations in Tacoma than at any other site in the country, at that time—some $662,000 between 1990 and 1993.[25]

The public relations campaign resonated to some degree in Ruston and Tacoma, and residents sometimes appeared confused as to how a Superfund cleanup would be funded, assuming that the EPA would use taxpayer dollars rather than making ASARCO pay. Loyalty to ASARCO and concern about the company declaring bankruptcy and what that might mean for its pension obligations and the town's future also influenced the local response to the EPA and Superfund. Some resident opposition to a Superfund cleanup of Ruston stemmed from a desire to return to normalcy—not to have one's town be designated a Superfund site, the object of stigma—and from a desire to have the contaminated smelter property quickly put back on tax rolls. These pressures are not uncommon in communities coping with legacy pollution and the prospect of a long-term environmental cleanup.

In Ruston, likely in part due to sophisticated public relations efforts, confusion over arsenic's health effects, and the tendency of some in the community to defend ASARCO no matter what, the EPA faced a community that was not uniformly convinced that cleanup was necessary and a company that was working to save money on this site and others across the country for which it was responsible. What emerged from this complex set of circumstances was a cleanup that was relatively inexpensive for ASARCO because the EPA used a cleanup standard that was tenfold less stringent than state law required and because the cleanup was limited geographically to properties within a mile of the smelter.[26]

But as the EPA saw it, some cleanup was better than none. Some at the EPA were concerned that ASARCO might declare bankruptcy, potentially leaving taxpayers with the entire bill.[27] The protectiveness of the EPA's cleanup in Ruston is easily challenged, particularly as more is learned about the toxicity and health effects of arsenic.

Although public health and environmental officials were aware that the smelter's contamination extended beyond the one-mile cleanup zone in Ruston, it was not until the late 1990s and early 2000s that the Tacoma smelter's complete soil contamination footprint was mapped. State and local agencies found arsenic and lead over state standards nearly everywhere they looked in the Puget Sound region, thanks to the tall stack—the industry's inexpensive solution to nearly a century worth of complaints about choking sulfur dioxide pollution. Arsenic and lead contamination has been documented extending throughout four counties.[28]

Today in Ruston, more than a decade of Superfund cleanup is coming to a close and a new luxury development is being built on top of the old smelter site. Many residents are ready to move on and put a toxic past behind them. But toxic metals remain in some yards, in slag, and buried in an on-site containment facility on the former smelter site that must be monitored in perpetuity. Considering the long history of concern over toxic metals in Ruston's environment and the evolving science of arsenic toxicity, the story is likely not over; health and environmental concerns are likely to resurface time and time again.

There is, however, a silver lining in the ASARCO Tacoma saga. In 2005 the company declared bankruptcy. Some believed this was a tactic to allow the company to escape its considerable environmental liabilities across the United States.[29] But in a rare and lucky break for communities long impacted by mining and smelter pollution, the bankruptcy settlement coincided with rising copper prices and put an unexpected $1.79 billion into government coffers for remediation of ASARCO sites across the country.[30] Washington State received $188 million, and some of the money will go toward remediating arsenic contamination that was not included in the EPA's limited Superfund cleanup of Ruston.[31] Even so, likely when the last of the cleanup dollars are spent, arsenic and other toxic metals will remain widely dispersed in the western Washington environment at concentrations far higher than background.

Roughly one-half billion dollars from the ASARCO bankruptcy settlement will go to the Silver Valley cleanup since ASARCO was one of the mining companies responsible for the valley's extensive environmental contamination. The Silver Valley cleanup is one of largest and most complex Superfund cleanups in the nation, projected to cost billions and span three decades. The cleanup is complex because the contaminated area is vast, the mines are among the deepest in the world, groundwater and surface water is polluted with mine waste, and heavy-metal-laden tailings are spread far and wide. Toxic metals are found in soils, house dust, the Coeur d'Alene River and its marshes and tributaries, and downstream in Lake Coeur d'Alene and the Spokane River. An estimated 120 million tons of mine tailings, including about 1.2 million tons of lead, contaminate the region. On a thirty-year timescale, the EPA is endeavoring to clean up a century's worth of damage.[32] Its aims are made more challenging by wind, rain, and flooding, which moves contamination through the valley.

Aside from the scale of the problem, the Silver Valley cleanup has also been complicated by contentious relationships between locals, state officials, and the EPA. Some Idaho political leaders have shown a disdain for the EPA, defended

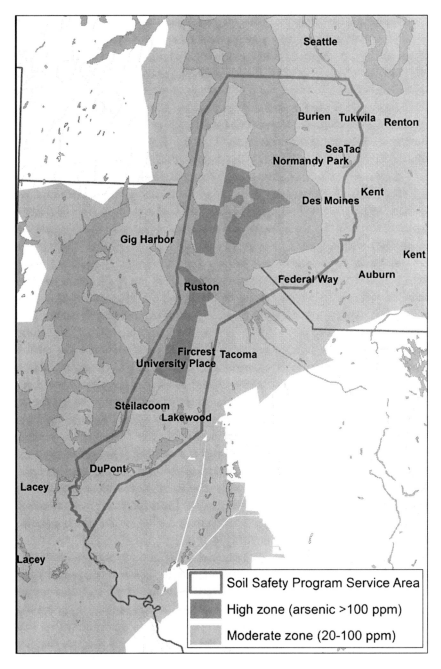

FIGURE 7.1 Arsenic soil contamination in the Puget Sound Region. The Soil Safety Program outline shows the area within which the Washington State Department of Ecology is working to decrease exposure to arsenic in soil through remediation and public education.

Used by permission of Washington State Department of Ecology.

the mining and smelting industries, disputed the need for cleanup, and sought to limit the EPA's reach—often by capitalizing on antigovernment sentiment aimed at the EPA—to try to limit the cleanup's scope. Some residents have also disputed the need for cleanup. Health advocates who favor strict cleanup have found themselves objects of community ire and have even feared for their personal safety. At times EPA staffers have resorted to using police protection at community meetings. Controversies have arisen at every turn, such as disagreements over expanding the boundaries of the site, how clean the soil should be, and even whether there should be any cleanup at all.[33]

In spite of the rancor, there have been notable successes. Blood lead levels have declined dramatically in the Silver Valley since the smelter's closure, and the EPA's remediation efforts (removing contaminated soils from over 2,500 yards) in the "Box," a twenty-one-square-mile area that includes Kellogg, has led to further reductions.[34]

Data from 2002 indicate substantial progress in reducing children's exposure to lead in the Silver Valley. Among 195 children under the age of ten who have been tested, the average blood lead level was 3.2 μg/dL. Although slightly higher than the U.S. average for 0 to 5 year olds, the average level is below CDC's recently revised level of concern (5 μg/dL). The highest blood lead level found was 21.3 μg/dL, which indicates the possibility of exposure above the level of concern and the importance of ongoing screening. In 2011 fourteen children from the Box were screened, and the highest blood lead level found was 5 μg/dL. However, in recent years fewer residents have opted to participate in annual blood lead screenings. For example, participation among Kellogg residents dropped from 195 children in 2002 to 8 in 2007.[35]

A recent study attributed the low rates of screening in Kellogg to the complex social history of the community, including shame stemming from lingering perceptions that having elevated lead levels can be attributed to poor parenting, a perception with deep roots going back to the 1970s. Other factors include a lack of support for testing from physicians and a belief that children are no longer at risk.[36]

Predictably, in recent years widespread residential soil contamination has been documented in El Paso, and cleanup of one thousand residential yards contaminated with lead and arsenic from emissions from the over eight-hundred-foot smokestack is ongoing. The ASARCO bankruptcy settlement has left about $50 million to clean up the 250-acre smelter site contaminated with arsenic, lead, cadmium, and other toxic metals as well as to fund the residential cleanup.[37]

However, when it comes to cleaning up such complex contaminated sites, the devil is in the details. Cleanup often amounts to simply moving contaminated dirt from yards or properties to landfills or other sites where the potential for recontamination of the environment continues. For example, arsenic- and lead-contaminated soils excavated from Ruston yards were moved to the smelter site to be used as the subbase for the cap over the smelter site. In the Silver Valley, huge repositories have been created to store contaminated soils and tailings, which must be monitored in perpetuity for their impact on groundwater and the environment.[38]

Taking a step back and examining what has been learned about community health effects from smelters since the 1970s, there are stark differences between Tacoma and El Paso and Bunker Hill. From Tacoma, little was learned about the public health effects of airborne arsenic due of a lack of investment in studying the issue both at the time exposure was occurring and subsequently. During the time the smelter was operating, there were no large-scale, well-designed studies that examined the health effects of arsenic on children. Notably, there have been no well-designed follow-up studies of children exposed to Tacoma's arsenic emissions in the 1970s, a time period when individual exposure measurements were taken, to attempt to discern long-term health effects.

In contrast, the childhood lead poisoning disasters near the Bunker Hill and El Paso smelters, tragic as they were, did advance the science of pediatric lead poisoning by helping to uncover the significance of subclinical lead poisoning in children, which contributed to setting an ambient standard for lead. The ambient standard and the phase out of lead in gasoline led to large reductions children's blood lead levels nationwide—a public health success story. This would not have been possible without the scientists, regulators, public health advocates, and community members who stood up to industry and challenged its right to pollute the air and poison children.

At least two of these scientists, Philip Landrigan and Ian von Lindern, went on to have distinguished careers in public and environmental health. Landrigan is considered by many the father of children's environmental health, responsible for helping to conceptualize children's unique vulnerability to environmental insults. Von Lindern is still working to protect children from lead poisoning by cleaning up the environment both in the United States, including the Silver Valley, and abroad.

As to what the long-term health impacts have been for children exposed to smelter pollution in the United States, there are few definitive answers. With the exception of a study of a small cohort of children exposed to high levels of lead near Bunker Hill in the early 1970s, there has been little systematic follow-up of U.S. communities exposed to smelter emissions.

The twenty-year follow-up study of Silver Valley children, conducted by the Agency for Toxic Substances and Disease Registry (ATSDR) (the agency responsible for health studies at Superfund sites), found that harm from childhood lead exposure has persisted in adults who were exposed as children after the baghouse fire. Those exposed to lead in childhood had "significant adverse central and peripheral neurological effects," including more neuropsychiatric symptoms than an unexposed comparison group. Another study found that babies born in the period following the baghouse fire near Bunker Hill had an increased risk of low birth weight at term and were more likely to be small for gestational age. Some Silver Valley residents who still struggle with physical and emotional health problems from exposure to lead in childhood have recently spoken out about the personal costs of living with the damage that lead can cause and the challenge of trying to cope with multiple health problems with few resources for medical care.[39]

In the late 1990s the ATSDR proposed screening adults exposed as children to Bunker Hill's lead for hypertension, renal disease, and adult attention deficit hyperactivity disorder—all possible sequelae of occupational or childhood lead exposure. The proposed screening would be offered to people who were heavily exposed to lead from the Bunker Hill smelter between 1973 and 1981. ASARCO and Hecla (mining companies involved in the cleanup of the Silver Valley) met with Idaho officials to voice their concerns and questioned the evidence linking lead with some of the health outcomes to be monitored. In the end the Idaho State Health Department decided it was "not in a position to successfully implement this ill-defined program" and declined to implement the program. Medical monitoring was apparently offered in Montana but not in Idaho.[40]

In El Paso the ATSDR funded a study of the incidence of multiple sclerosis (MS) in people who grew up in Kern Place and Smeltertown and found approximately twice the expected rate of the disease. Although the researchers were careful not to attribute the observed MS cases to exposure to smelter emissions, a purpose of the study was to explore the hypothesis that environmental heavy metal exposure may play a role in the etiology of neurological diseases such as MS.[41] There have not been long-term, comprehensive follow-up studies of children from Smeltertown, the section of El Paso where children were most highly exposed.

Although industry and government largely failed to protect communities from smelter pollution throughout the century, some community members won redress in the courts. In both Tacoma and Bunker Hill, residents won damages against smelting companies. These were difficult cases, and the fact that they were won is probably largely attributable to the intellect and tenacity of the plaintiffs' lawyers along with some lucky breaks. In Tacoma plaintiffs won $67.5 million from ASARCO for property damage, and at Bunker Hill children with health damage won approximately $30 million. While significant for the individuals involved, this type of compensation falls short of a public health solution that would aim to provide benefit to all through concerted efforts to reduce exposure and through medical monitoring and treatment with publicly reportable results that would advance the state of knowledge regarding the long-term sequelae of exposure to toxic metals in the environment. The courts are a last-ditch solution when government abdicates its regulatory responsibility.

In both cases, internal industry documents obtained through discovery were sealed as part of settlements. Internal Bunker Hill documents only became publicly available because the EPA petitioned to access them as part of the Superfund process to establish Gulf's liability. With respect to Tacoma, the majority of internal ASARCO documents obtained during discovery remain sealed. A part of the settlement included some medical monitoring of the Ruston and North Tacoma population, but the results are apparently not publicly available.

Looking back on nearly a century of smelting and conflicts over smelter pollution in the United States, limiting smelter pollution has been framed as a choice between health and jobs and economic prosperity, with the jobs argument typically winning out. Industry often has had to have its feet held to the fire to make even small improvements, but regulators have done this only reluctantly. Pressure for jobs and short-term economic gains have left behind communities devastated by toxic waste and facing environmental cleanups costing tens of millions of dollars or more. Both government and industry have tended to respond to harms that should have been anticipated in a way that is reactive rather than proactive. Even today, near the Hayden smelter in Arizona a school complex is within fifteen hundred feet, the public park and swimming pool sit at the fenceline, and an ore conveyor runs near residential properties.[42]

In retrospect, it is difficult to conclude that the long timeline for setting federal environmental standards for smelters and limited enforcement (essentially the tacit acceptance of chronic pollution from some smelters) was accidental or exceptional, particularly since a comparable industry—electric utilities, which could not move overseas—has only recently had limits placed on toxic mercury and other hazardous air pollutants from old coal-fired power plants. That industry is still years away from full compliance.[43]

In the case of primary nonferrous smelters, it appears that in many cases the EPA made a strategic decision to allow them to run out the clock on their operations rather than take on the political struggle and endure the fallout necessary to set (in the case of arsenic) and enforce (in the case of lead) environmental regulations. Even in the 1970s the agency knew that many of the worst polluting smelters were nearing the end of their useful life spans due to obsolescent facilities and changing markets. By the 1980s, though only ten years old, the EPA was facing an intense antiregulatory backlash and did not want to take the blame for smelter closures. All the industry had to do was buy time and keep up its offensive against regulation and the EPA until the facilities were no longer needed. Unfortunately, the health of people who lived in these communities was sacrificed while the industry played out its hand.

The smelters that remained open in the United States into the 1990s and beyond did reduce their sulfur dioxide and heavy metal pollution considerably over 1970 levels. These improvements are largely attributable to challenges to industry power by communities, scientists, and public officials. Without a change in public values, citizen organizing, advocacy, and independent science, smelters would likely still be pumping out enormous quantities of pollution.

But U.S. communities that are home to primary nonferrous smelters still live with both legacy and ongoing pollution even though substantial improvements have been made. In Herculaneum, Missouri, the site of the country's last operating primary lead smelter, that smelter put nearly twenty tons of lead in the air in 2008, making it the largest point source for lead in the United States. It was only in 2008, some thirty years after the standard was set, that the EPA agreed that the air around the smelter appeared to be in attainment with the 1978 lead standard.[44]

In the early 2000s public health officials discovered that almost half of the children living closest to the Herculaneum smelter had blood lead levels over 10 μg/dL. Buyouts of property owners ensued, and families were relocated. Fortunately, most recent tests indicate that no children have blood lead levels exceeding 10 μg/dL. But lead emissions continue to contaminate the environment—some yards that were cleaned up have become recontaminated with lead in the space of a decade.[45] The EPA recently tightened the federal ambient lead standard (0.15 μg/m^3), and the Doe Run Company announced plans to close Herculaneum. When the plant closes in 2013, it will mark the end of primary lead smelting in the United States.

In Hayden, Arizona, a largely Mexican American community lives in the shadow of an ASARCO copper smelter. Recent EPA figures put concentrations of arsenic in the air in Hayden sixty times higher than in areas not impacted by smelting. Cadmium, lead, chromium, and copper also contaminate Hayden's air. A recent risk assessment estimated that Hayden residents are at an increased risk of cancer, largely from exposure to arsenic in air.[46]

In the Salt Lake City metropolitan area, Kennecott operates an open-pit mine, smelter, and refinery, which public health and parent activists blame in large measure for the region's air pollution problems. In 2010 the smelter and refinery reported 32 million pounds of toxic releases, ranking twelfth highest in the United States.[47] A recent lawsuit brought by health, parent, and environmental groups alleges that the company is violating the Clean Air Act and exposing the community to ongoing particulate air pollution. One physician compared air pollution in the region to "breathing second-hand smoke" and, reminiscent of some Tacoma activists from the 1980s, said, "It is time for citizens to take matters into their own hands."[48]

Looking beyond the smelting industry, cleaning up hazardous air pollution (for example, mercury, arsenic, cadmium, benzene, and other highly toxic air contaminants) in the United States has proved to be a daunting task. In the first twenty years after the passage of the 1970 Clean Air Act very little progress was made in cleaning up air toxics. Many observers blamed the approach Congress took in the act, which directed the EPA to set health-based standards for hazardous air pollutants without considering the cost to industry, and provided little guidance to regulators on how to do so. The EPA did little to address air toxics between 1970 and 1990, and the public missed out on twenty years of regulatory efforts to limit toxic exposures from air.

By the early 1980s some in Congress were pushing for a renewed emphasis on the cost to industry to control air toxics in regulatory decision making. The 1990 revisions to the Clean Air Act, passed under the George H. W. Bush administration, abandoned the use of strict health-based standards for air toxics and ushered in a two-phased approach in which the cost of regulation is prominently considered. In the first phase, industry is to install maximum achievable control technology (MACT), determined by the EPA, after considering the costs of compliance. After MACT, residual risks to human health are to be assessed. If risks remain, further reductions in emissions are to be made to protect public health. The revisions were heralded with promises that the country would finally succeed in cleaning up hazardous air pollution.

But as the smelting story shows, cleaning up hazardous air pollutants is not as simple as having laws on the books. Achieving clean air takes political will, commitment to implementing laws, adequate funding, and enforcement. States have primary responsibility for reducing air toxics, and there is significant variability by state. Ongoing clean up challenges are attributable to the lack of funding, political will, regulatory oversight, and enforcement.

It is difficult to know for certain how effective regulators have been in reducing hazardous air pollution since 1990 due to the limitations of existing data sources. However, the EPA estimates that there has been a 42 percent reduction in hazardous air pollution emissions between 1990 and 2005.[49] Although the EPA's regulatory efforts are probably responsible for a significant proportion of the estimated reductions, other factors may also be contributing, such as changes in reporting, industry closures, or decreased production.

Despite the reductions, air toxics are still a pressing public health issue. The EPA estimates that all Americans have an increased cancer risk greater than 10 in 1 million from exposure to air toxics, and some communities face relatively high risks.[50] A recent Center for Public Integrity and National Public Radio series, *Poisoned Places*, investigated the ongoing problem of community exposure to hazardous air pollution. Communities with cement kilns, coke plants, and smelters (such as Hayden, Arizona) were profiled.[51] These communities face many of the same problems that smelting communities faced in the 1970s when trying to get relief from chronic pollution, including the lack of data on actual pollution levels, reluctant regulators, long timelines for regulatory action, few penalties for polluters, and the need to organize politically and turn to the courts for redress. Low-income communities and communities of color are often disproportionately burdened by chronic industrial air pollution.

In the ongoing struggle to clean up hazardous air pollution, industry concerns about the costs of pollution control still bog down the debate, which suggests the ongoing influence of industry in shaping the actions of regulators and elected officials. Cost arguments made by industry are compelling to regulators and government officials, particularly during economic downturns, and can only be effectively counteracted by an engaged public that is no longer willing to accept the costs borne by society from failing to control pollution. For example, the EPA estimates that the annual value of the health benefits of reducing mercury and air toxics will be between $37 billion and $90 billion. Direct benefits of the 1990 Clean Air Act are estimated to reach $2 trillion by 2020, and the cost of compliance in the same year is estimated at a fraction of this, at $65 billion.[52]

Since the 1970s the overall trend in the United States is toward improvement in air quality, but in many cases gains have taken a long time and have not been uniformly protective. The public health and environmental impacts of air pollution remain significant, with climate change topping the list. Research on the health effects of air pollution has become increasingly sophisticated and leaves little doubt about its pervasive and far-reaching consequences for human health, including adverse effects on infant development, premature mortality, respiratory disease, heart disease, and cancer. As a nation we have accepted incremental and uneven improvement in the air we breathe, signaling that political power is still wielded disproportionately by polluters rather than those most affected by pollution.

Today a similar story of endangerment to public health from nonferrous smelting unfolds overseas, where the primary smelting industry largely operates. China, India, Russia, Chile, and Zambia are now significant producers of nonferrous metals, and predictable and largely preventable environmental and public health problems are found near smelters abroad.[53] For example, near a defunct lead smelter in Kabwe, Zambia, unregulated mining and smelting have resulted in average BLLs in children living nearby of between 50 and 100 µg/dL, with some children's as high as 200 µg/dL. In Norilsk, Russia, a city that is heavily polluted by mining and smelting, a metals smelting complex discharges over one thousand tons of toxic metals into the air each year. Sulfur dioxide pollution has killed forests nearby and residents complain of health problems.[54] A recent study of sulfur dioxide pollution from smelters in Peru and Russia found that "these point sources are still a dominant source of anthropogenic sulfur dioxide emissions in their region."[55] A 2007 report on the top ten most polluted places in the world found that half were mining and or smelting sites.[56] In La Oroya, Peru, around a multimetal smelter, in the early 2000s nearly all children living in the surrounding town who were tested had elevated blood lead levels.[57] Multiple accounts of childhood lead poisonings near smelters have recently emerged from China. Human Rights Watch has reported that industrial lead poisoning in children in some Chinese communities is being covered up.[58]

That the industry continues to harm health and the environment, inflicting predictable and preventable harm on new generations of children, is a depressing testament to ongoing industry and government indifference to the significant environmental and public health consequences of smelting.

Conclusion

> "At the moment I heard the doctor say that, my heart was shattered . . . we wanted this child to have everything. That's why we worked this hard. That's why we poisoned ourselves at this factory. Now it turns out the child is poisoned too. I have no words to describe how I feel."
>
> −Battery factory worker, Mr. Han, on being told by a doctor that his three-year-old daughter has lead poisoning. The family lives across the street from the battery factory in China.[1]

Forty years after the discoveries at El Paso and Bunker Hill, half a world away, parents are still coping with the same anguish, their children poisoned, perhaps impaired for life because of a lack of industry and government commitment to protecting the health of people and the environment. If the U.S. experience of smelting and its community health consequences over the twentieth century holds lessons for today, it makes clear that protecting health from industrial pollution requires an engaged and educated citizenry, funding for independent research, an alert and proactive public health community, and strong public and government commitment to health and environmental protection.

Today community environmental health struggles continue both in the United States and around the world, and they often share the same basic outlines of the smelter story. Community concerns about health impacts of industrial pollution are often met with industry claims that health effects are uncertain, unproven, or overstated. Too often using scientific uncertainty as an excuse for inaction, regulators stand on the sidelines of these controversies, reluctant to stake out a strong position in favor of precaution.

Unless the exposure of concern has been well studied (for example, lead) and unless there are clear public health or environmental standards that are being violated, debates over exposure and harm can drag on for decades. Proving harm to community health through epidemiologic studies can be challenging, and answers do not come quickly. The most definitive epidemiologic studies, which examine health effects prospectively, may take decades to complete and cannot yield conclusive results when conducted in small populations.

The smelting story should read as a cautionary tale as we encounter new threats to community health, whose proponents argue that health effects have not been proven. For all of the doubts raised by the smelting industry about the impact of arsenic, lead, and sulfur dioxide on human health in the 1970s and earlier, further research has only borne out and amplified the concerns of those who took precautionary positions early on. The list of cancers and chronic diseases with which arsenic is now associated has only grown, and exposure to lead in childhood is increasingly understood to be a lifelong health hazard, with no safe level of exposure. Histories of twentieth-century toxic exposures and community health disasters make a strong case for precaution in the face of unknown health effects.

Over the course of the twentieth century the "success" of the western smelting industry in substantially dodging regulation and externalizing the costs of pollution onto surrounding communities was based largely on three intertwined narratives, which continue to be common themes in battles over community environmental health.

The first narrative was the notion that Americans could not have jobs, health, *and* a clean environment—we would have to choose. Coupled with threats to close if health concerns were prioritized, ASARCO, in particular, relied on this argument for much of the twentieth century.

The second strand of the argument was that of introducing doubt—was health really being harmed by smelter emissions? The question had significant regulatory and liability implications. For many years the industry succeeded in using science for dominating the narrative and downplaying concerns about adverse health and environmental effects. Resistance to the industry narrative originated from affected communities and from some researchers and set the stage for independent investigations and eventual regulation. The timelines for protecting health and the environment, however, were long. Even after 1970, as more independent research lent credibility to concerns about the health and environmental impacts of smelting, smelting companies continued to fund self-interested science to keep the doubt narrative alive and prolong debates over regulation.

The third narrative was of "overregulation" or government intrusion marked by appeals to independence, individualism, and antigovernment sentiment. Government's interest in protecting health and the environment was characterized as dangerous overreaching that threatened our economy and our freedom. Industry successfully promoted this argument, playing on core American values. Sophisticated public relations efforts helped to unify these arguments in a compelling package to influence policymakers and the public. At times regulators embraced these narratives so that conflicts with industry and further regulation or enforcement could be avoided.

The central question that the smelting story raises for the twenty-first century is can we do better? Can we pursue a path that is more sustainable? Can we have a more equitable distribution of power between government, industry, and people? Can we place more value on public health and sustainability? Can industry behave more responsibly? Can we take a precautionary approach, erring on the side of caution rather than waiting for illnesses, injuries, or deaths to mount? Can the field of public health do a better job of protecting people from industrial pollution and environmental health threats?

Forty years after the establishment of the EPA and federal environmental regulatory authority, the environmental health challenges we face are daunting. Despite progress, air pollution remains the most pressing environmental issue of our time. Effective regulatory action on climate change is stymied by the same kinds of strategies used by the smelting industry in the twentieth century: casting doubt on the science, using public relations techniques to shape public opinion, and claiming that regulation will harm the economy and cost jobs. Recent regulatory and industry failures left a Gulf region devastated by a massive oil spill, and industrial emissions at home and overseas continue to threaten children's health. Twenty-first-century scientific understandings make clear the far-reaching ecological consequences of industrial pollution as toxic metals and other types of industrial pollution from Asia impact air quality in the western United States and as Fukushima's radiation circled the globe. What were previously thought of as local nuisances are global problems that threaten human health and the life-sustaining systems of our planet. Still, effective regulation of many significant environmental health problems remains elusive.

Part of the challenge is the difficulty political and government leaders have acting in the public interest due to the ongoing undue influence industry has on them within regulatory agencies and at all levels of government. Even the most well meaning and public-interest-oriented EPA administrators and state environmental officials face extraordinary obstacles to challenging industry power, and their efforts may be undermined by Congress, the president, or other White House offices such as the Office of Management and Budget, or at the state level by governors and state legislatures. Too often it is easier for regulators to study problems, put off regulation, and wait for environmental groups to challenge agencies in court to enforce existing laws, rather than to tackle challenges head-on.[2]

Another challenge to protecting health and the environment is our hyperpartisan political discourse. For some, the EPA has become the most reviled institution in government, a favorite target of attack, with some politicians calling it a "job-killing agency," arguing against any role for the agency in controlling greenhouse gas emissions and calling for a halt to environmental regulation in the service of economic growth.[3] Although these views do not reflect those of most Americans, they have undeniably become a part of our political discourse and make decisive regulatory action more difficult.

In an era when criticisms of government's size, power, and role dominate politics, the smelter story is an instructive case study of what can happen in the face of weak and ineffective government regulation and with the prioritization of industry interests above the health of children, workers, and the environment. For too many decades many western smelters polluted with almost total impunity. The costs are largely incalculable and have been borne by workers and the public.

As historians and public health scholars scrutinize environmental and public health conflicts over the course of the twentieth century, lessons for the struggle to achieve health are becoming clearer. Western smelting companies are not unique. Many other industries have shown indifference to the human health and environmental consequences of their products and production. For example, the relentless pursuit of profits by the tobacco industry has been well documented, and in recent decades the industry has moved aggressively into the developing world, marketing its product to billions with few restrictions.[4] The lead industry denied and covered up the harms of its deadly product for decades, harming untold numbers of children and workers.[5]

The costs to individuals and to society from such indifference makes a powerful argument for a renewed, strong, and consistent investment in the system of governmental public health and environmental protection that aims to insulate policymakers from undue industry influence and provide them with some measure of independence from political interference.

The smelting story exposes the consequences of largely unchallenged industry power for the health of communities. Preventing and solving environmental and public health problems requires a strong commitment to democratic values, far-sighted thinking, an educated public, independent science, effective organizing, a pervasive ethic of sustainability, and a valuing of ends other than industry profits. We must find a better way in the twenty-first century as the challenges we face are even bigger.

Notes

INTRODUCTION

1. Kery Murakami, "Tumbling Landmark—Smokestack Bites Dust," *Seattle Times*, January 18, 1993, A-1. Details on the stack demolition reconstructed from ibid.; Rob Taylor, "Smokestack Now Smithereens; Smokestack's Demolition Part of Toxic Cleanup," *Seattle Post-Intelligencer*, January 18, 1993, B-1; "ASARCO Stack Demolition Tacoma," YouTube video, http://www.youtube.com/watch?v=X6YPDbWilCo, accessed January 8, 2013; and Bill Tobin, interview with author August 5, 2005, Vashon Island, Washington. Digital audio recording. File in author's possession.
2. Donald MacMillan, *Smoke Wars: Anaconda Copper, Montana Air Pollution, and the Courts, 1890–1924* (Helena: Montana Historical Society Press, 2000); and Timothy J. LeCain, *Mass Destruction: The Men and Giant Mines That Wired America and Scarred the Planet* (New Brunswick, NJ: Rutgers University Press, 2009).
3. Irwin Unger and Debi Unger, *The Guggenheims: A Family History* (New York: HarperCollins, 2005), 165, 287; Isaac F. Marcosson, *Metal Magic: The Story of the American Smelting and Refining Company* (New York: Farrar, Straus and Company, 1949); and Michael Malone, "The Collapse of Western Metal Mining: An Historical Epitaph," *Pacific Historical Review* 55 (August 1986): 455–464.
4. See, for example, MacMillan, *Smoke Wars*; M. A. Church, "Smoke Farming: Smelting and Agricultural Reform in Utah, 1900–1945," *Utah Historical Quarterly* 72 (Summer 2004): 196–218; and John D. Wirth, *Smelter Smoke in North America: The Politics of Transborder Pollution* (Lawrence: University Press of Kansas, 2000).
5. See, for example, George R. Hill, "Report: Condition of Vegetation at Tacoma, August 26–27, 1952," *Branin v. ASARCO*; S. M. Lane, Manager, East Helena Plant to W. G. Rouillard, "Memorandum: Horses—East Helena," August 11, 1965, *Branin v. ASARCO*, 93-CV-5132, Accession # 021-04-0102, Box 1-9, Location # 3050865, settlement papers obtained at Seattle Federal Records Center, hereafter referred to as *Branin v. ASARCO*. Unknown author, "Handwritten notes: WJB Smith—Dogs," April 11–April 15 1969, *Branin v. ASARCO*.
6. See Washington State Department of Ecology, "Living with a Toxic Legacy," http://www.ecy.wa.gov/programs/tcp/sites_brochure/tacoma_smelter/2011/ts-hp.htm, accessed January 13, 2013.

7. "Human Lead Absorption—Texas" *Morbidity and Mortality Weekly Report* (December 8, 1973), 405–407, http://www.cdc.gov/mmwr/preview/mmwrhtml/lmrk095.htm, accessed April 24, 2013.
8. See Gene Baker, "Confidential: Events Pertaining to Lead Health Problem in Kellogg" (undated, reported to have been written in collaboration with attorneys in 1979 or 1980), in *Edna Grace Yoss, Raymond Hans Yoss, Arlene Mae Yoss, Richard A. McCartney, Christina M. McCartney, Paula A. McCartney, Raymond E. Dennis, and Harley Dennis, by their guardian ad litem, Brain J. Linn, Plaintiffs, v. Bunker Hill Company, a Delaware corporation; Gulf Resources & Chemical Corporation, a Delaware corporation; Defendants*, No. CIV-77–2030, United States District Court for the District of Idaho (hereafter *Yoss et al. v. Bunker Hill and Gulf*); Bunker Hill Company, Lead Smelter Emission Record, 1971–1977 (n.d.), *Yoss et al. v. Bunker Hill and Gulf*, Exhibit #32; Philip J. Landrigan to the Record, "Memorandum: Lead Emission Patterns—Kellogg, Idaho, Smelter," July 18, 1975, *Yoss et al. v. Bunker Hill and Gulf*, Files of Paul Whelan, Plaintiff's attorney.
9. P. J. Landrigan, E. L. Baker Jr., R. G. Feldman, D. H. Cox, K. Eden, W. Orenstein, J. Mather, A. Yankel, and I. Lindern, "Increased Lead Absorption with Anemia and Slowed Nerve Conduction in Children Near a Lead Smelter," *Journal of Pediatrics* 89 (1976): 904–910.
10. Samuel Milham Jr., MD, and Terrance Strong to Wallace Lane, MD, Washington State Health Services Division, "Memorandum: Tacoma Smelter Study," November 2, 1972, Environmental Protection Agency Site File, Commencement Bay/Nearshore Tideflats ASARCO Smelter Facility Site File (hereafter, ASDSF), 1.1.1.
11. David C. Bellinger, "Lead," *Pediatrics* 113, Suppl. 4 (2004): 1016–1022.
12. MacMillan, *Smoke Wars*; and Wirth, *Smelter Smoke in North America*.
13. MacMillan, *Smoke Wars*; Church, "Smoke Farming," 196–218; and Wirth, *Smelter Smoke in North America*.
14. The El Paso study conducted by James McNeil was funded in equal thirds by ILZRO, ASARCO, and the Ethyl Corporation (producer of tetraethyl lead, the gasoline additive). See John J. Sheehy to Files, "Memorandum: Gulf Resources & Chemical Corporation—ILZRO," October 2, 1974, *Yoss et al. v. Bunker Hill and Gulf*. In northern Idaho Bunker Hill initially turned to ILZRO to conduct a study of children's health in the Silver Valley, but there was opposition to this. Instead, Bunker Hill funded the Shoshone Project, which was said to be an independent study but the company provided the cash contributions for the study. See "Memorandum of Agreement between J. H. Halley, Dr. James A. Bax, and Glen Wegner," *Yoss et al. v. Bunker Hill and Gulf*. ILZRO also remained involved in the Shoshone Project. See, for example, Gene M. Baker to Jerome F. Cole, "Letter," June 18, 1975, *Yoss et al. v. Bunker Hill and Gulf*, Files of Paul Whelan, Plaintiff's attorney. Baker asks for Cole's assistance in analyzing a study of Kellogg children's intelligence tests that was part of the Shoshone Project and promises to forward to ILZRO the names of the Kellogg children who had taken part in the tests.

15. David Michaels, *Doubt Is Their Product: How Industry's Assault on Science Threatens Your Health* (New York: Oxford University Press, 2008).
16. See, for example, James D. Callaghan, Hill & Knowlton, Inc. to Philip E. Robinson, Lead Industries Association, Inc., "Memorandum: Further Distribution of the McNeil Study," July 1, 1974, *Yoss et al. v. Bunker Hill and Gulf*, Files of Paul Whelan, Plaintiff's attorney; James L. McNeil, "Evaluation of Long Term Effects of Elevated Blood Lead in Asymptomatic Children," Proposal to the International Lead Zinc Research Organization, August 1972 (rev. November 1972), *Yoss et al. v. Bunker Hill and Gulf*, Files of Paul Whelan, Plaintiff's attorney; James D. Callaghan, Hill & Knowlton, Inc., to Philip E.

Robinson, Lead Industries Association, Inc., "Memorandum: Idaho Lead Smelter Story," September 6, 1974, *Yoss et al. v. Bunker Hill and Gulf*, Files of Paul Whelan, Plaintiff's attorney; Dale Henderson, Inc. Public Relations & Marketing, "Confidential Report" (n.d.), *Yoss et al. v. Bunker Hill and Gulf*; H. Dale Henderson to Mr. Gene M. Baker, "Letter," August 21, 1975, *Yoss et al. v. Bunker Hill and Gulf*, Files of Paul Whelan, Plaintiff's attorney; John P. Sammons, Senior Vice President, Hill and Knowlton, to Lawrence W. Lindquist, Manager ASARCO Tacoma Plant, July 29, 1983, *Branin v. ASARCO*; and Mundy & Associates, "Public Relations Expenditures (Elgin Seyferd)" (report excerpt, n.d.), *Branin v. ASARCO*.

17. See, for example, "Pollution Rules May Force Asarco to Shut Smelter," *Chemical Week,* October 29, 1975; "Emission Fines Sought," *Chemical Week,* March 3, 1976; Robert Lindsey, "Hard Times in Copper Country," *New York Times,* December 12, 1976, F1; "EPA Rule Controlling Lead Levels in Air Is Made Final But It May Be Altered Later," *Wall Street Journal,* May 31, 1978, 4; "EPA Lead Emission Rules Can't Be Met," *Coeur d'Alene Press,* June 8, 1979; and "U.S. Must Face Minerals Issues Now: ASARCO's Barber Tells Strategic Resources Conference," *Engineering and Mining Journal,* January 1982, 9.

18. ASARCO Inc., "News Release," May 15, 1972, KCA, Box 112 Health Department Director Issues Files, Folder 11 Lead Poisoning 1972–1979. In its news release on the consent decree entered into with the City of El Paso and the State of Texas, ASARCO asserted that "the current operation of the plant is not endangering the health of El Paso residents. Publicity and actions by city officials during the trial implied that the health of the entire City of El Paso was endangered. The medical examinations and therapy called for in Article IV of the decree were voluntarily initiated by Asarco in February when the elevated blood lead levels in Smeltertown children, apparently due to contaminated soil, were first discovered." With respect to Bunker Hill, see, for example, Gene M. Baker, "The Bunker Hill Company-Environmental Control Update," December 7, 1974. Paper presented at the Northwest Mining Association Convention at Spokane, Washington, *Yoss et al. v. Bunker Hill and Gulf*; State of California, Air Resources Board, "In the Matter of: Reconsideration of the Ambient Air Quality Standard for Lead, Comments of Ronald K. Panke," November 3 and 4, 1975, *Yoss et al. v. Bunker Hill and Gulf*; Ian H. von Lindern, "Letter: Re: Ambient Air Standard Docket 77-1," May 24, 1980, *Yoss et al. v. Bunker Hill and Gulf*, Files of Paul Whelan, Plaintiff's attorney. With respect to Tacoma, see Testimony of Dr. Ted A. Loomis, attachment to C. J. Newlands to R. J. Muth, February 16, 1973, ASDSF 1.6.3; Armand Labbe, "Presentation before the Board of Directors of the Puget Sound Air Pollution Control Agency at the Public Hearing for Proposed Amendments to Regulation I for Controlling the Emission of Arsenic," February 14, 1973, ASDSF 13.5; Herbert G. Lawson, "Dilemma in Tacoma: Smelter's Emissions Threaten Populace, But So Does Possible Loss of 1,000 Jobs," *Wall Street Journal,* July 16, 1975, 36; and "Smelter Promises to Keep Trying," *Vashon Maury Island Beachcomber,* January 22, 1976.

19. See for example, Allen M. Brandt, *The Cigarette Century: The Rise, Fall, and Deadly Persistence of the Product That Defined America* (New York: Basic Books, 2007).

CHAPTER 1 THE TACOMA SMELTER

1. "Asarco Fallout Casts Long Shadow; New Research: Studies Find That Copper Smelter Caused Broader Contamination Than Suspected, Also That Arsenic May Be More Toxic," *Tacoma News Tribune,* June 2, 2002, A1.

2. Gregory L. Glass, *Credible Evidence Report: The ASARCO Tacoma Smelter and Regional Soil Contamination in Puget Sound, Final Report*, September 2003, Prepared for Tacoma-Pierce County Health Department and Washington State Department of Ecology, http://www.ecy.wa.gov/programs/tcp/sites_brochure/tacoma_smelter/Sources/Credible_Evidence/web%20pieces/X_cred%20entire%20report.pdf.

3. Murray Morgan, *Puget's Sound: A Narrative of Early Tacoma and the Southern Sound* (Seattle: University of Washington Press, 1979), 306–309.

4. "Ruston More Than Doubles Population and Assessed Valuation in 18 Months; Smelter City Is Prospering," *Tacoma Daily Ledger*, June 14, 1908, 22.

5. Ibid.

6. Isaac F. Marcosson, *Metal Magic: The Story of the American Smelting and Refining Company* (New York: Farrar, Straus, 1949); and Irwin Unger and Debi Unger, *The Guggenheims: A Family History* (New York: HarperCollins, 2005), 165.

7. Marcosson, *Metal Magic*, 282.

8. Robert Coughlin to Randy Smith, Region X EPA, "Report: ASARCO Closure," January 7, 1985, obtained from Washington State Department of Ecology, ASARCO electronic files, hereafter referred to as Ecology electronic files.

9. See, for example, John Burnett, "A Toxic Century: Mining Giant Must Clean Up Mess," *National Public Radio*, February 4, 2010, http://www.npr.org/templates/story/story.php?storyId=122779177; and Bankruptcy Creditors' Service, Inc., *ASARCO Bankruptcy News* 1 (August 11, 2005).

10. Donald MacMillan, *Smoke Wars: Anaconda Copper, Montana Air Pollution, and the Courts, 1890–1924* (Helena: Montana Historical Society Press, 2000); and K. Aiken, "Western Smelters and the Problem of Smelter Smoke," in *Northwest Lands, Northwest Peoples: Readings in Environmental History*, ed. D. D. Goble and P. W. Hirt, 502–522 (Seattle: University of Washington Press, 1999).

11. Utah Department, General Manager's Office [name illegible] to Mr. Karl Eilers, Vice President, American Smelting and Refining Company, July 27, 1916, *Branin v. ASARCO*; and L. V. Olson, Director Department of Agricultural Research, to D. A. Somerville, Tacoma Smelter, February 7, 1951, *Branin v ASARCO*.

12. Environmental Protection Agency, APTI Virtual Classroom, "Lesson 16: Nonferrous Smelters," http://yosemite.epa.gov/oaqps/EOGtrain.nsf/DisplayView/SI_431_16?OpenDocument.

13. World Bank Group, *Pollution Prevention and Abatement Handbook, 1998: Toward Cleaner Production* (Washington, DC: World Bank, 1999), 292; EPA, "Integrated Science Assessment for Sulfur Oxides-Health Criteria," 73 Fed. Reg. 178 (September 12, 2008), 53002–53003, http://www.epa.gov/fedrgstr/EPA-AIR/2008/September/Day-12/a21317.pdf; and C. H. Fulton, *Metallurgical Smoke*, Bureau of Mines, Bulletin 84 (Washington, DC: Government Printing Office, 1915).

14. World Bank Group, *Pollution Prevention and Abatement Handbook*, 292.

15. Washington State Department of Ecology, *B & L Woodwaste Landfill*, https://fortress.wa.gov/ecy/gsp/Sitepage.aspx?csid=2297, accessed June 4, 2013; and EPA Region 10, *EPA Record of Decision: Commencement Bay/Nearshore Tideflats, OU 2, ASARCO Tacoma Smelter Facility*, March 24, 1995, EPA/ROD/R10-95/122, http://www.epa.gov/superfund/sites/rods/fulltext/r1095122.pdf.

16. Environmental Protection Agency, APTI Virtual Classroom, Lesson 16, Nonferrous Smelters. http://yosemite.epa.gov/oaqps/EOGtrain.nsf/fabbfcfe2fc93dac85256afe00483cc4/e75c917eecd8575685256b88004e6f05/$FILE/si431-lesson16.pdf, accessed June 3, 2013.

17. Ibid.; and ASARCO Incorporated, *Compliments of ASARCO Incorporated, Tacoma Plant* (n.d.), Washington State Historical Society, ASARCO Collection, Box 1, File 6. This is an undated history of the Tacoma plant produced by ASARCO.
18. S. Hong, J. Candelone, C. C. Patterson, and C. F. Boutron, "History of Ancient Copper Smelting Pollution during Roman and Medieval Times Recorded in Greenland Ice," *Science* 272 (1996): 246–249.
19. Robert E. Swain, "Smoke and Fume Investigations: A Historical Review," *Industrial and Engineering Chemistry* 41 (1949): 2384–2388; and Aiken, "Western Smelters," 502–522.
20. MacMillan, *Smoke Wars*, 98; and H. Galbraith, K. LeJeune, and J. Lipton, "Metal and Arsenic Impacts to Soils, Vegetation Communities, and Wildlife Habitat in Southwestern Montana Uplands Contaminated by Smelter Emissions I. Field Evaluation," *Environmental Toxicology and Chemistry* 14 (1995): 1895–1903.
21. Michael Malone, "The Collapse of Western Metal Mining: An Historical Epitaph," *Pacific Historical Review* 55 (August 1986): 455–464; and MacMillan, *Smoke Wars*, esp. 217.
22. MacMillan, *Smoke Wars*, 88–89.
23. J. K. Haywood, "Injury to Vegetation and Animal Life by Smelter Fumes," *Proceedings of the American Chemical Society*, vol. 29 (Easton, PA: The Chemical Publishing Company, 1907), 998–1009.
24. See MacMillan, *Smoke Wars*, for a detailed account of the conflicts between the Deer Lodge Valley farmers and the Anaconda Company. ASARCO's dispute with farmers in the Salt Lake Valley is discussed in M. A. Church, "Smoke Farming: Smelting and Agricultural Reform in Utah, 1900–1945," *Utah Historical Quarterly* 72 (Summer 2004): 196–218.
25. MacMillan, *Smoke Wars*; and Church, "Smoke Farming," 196–218.
26. MacMillan, *Smoke Wars*, 91.
27. Ibid., 90.
28. "Mammoth New Chimney Just Completed at Tacoma Smelter," *Daily Ledger,* May 7, 1905, 4.
29. "Mammoth New Chimney," 4.
30. W. D. Harkins and R. E. Swain, "The Determination of Arsenic and Other Solid Constituents of Smelter Smoke, with a Study of the Effects of High Stacks and Large Condensing Flues, Papers on Smelter Smoke: First Paper," *Proceedings of the American Chemical Society*, vol. 29 (Easton, PA: The Chemical Publishing Company, 1907), 970–998, at 996.
31. Michael O. Varner, *ASARCO Incorporated: The Department of Environmental Sciences* (n.d.), *Branin v. ASARCO*, 93-CV-5132, Accession # 021-04-0102, Box 1-9, Location # 3050865, settlement papers obtained at Seattle Federal Records Center, hereafter referred to as *Branin v. ASARCO*.
32. See, for example, Utah Department, General Manager's Office [name illegible] to Mr. Karl Eilers, Vice President, American Smelting and Refining Company, July 27, 1916, *Branin v. ASARCO*. In this internal ASARCO letter, the manager of the Utah department, responsible for air pollution studies at the company's demonstration farm, writes Vice President Eilers that "in regard to the handling of smoke matters; this should be entrusted to people who have a knowledge of it, and who have had the experience, to advise you properly. We have built up, in the Utah Department, an organization for that purpose, which should be of great benefit to the company's interests, in the handling of such matters."
33. For more on ASARCO's role in research on sulfur dioxide benefitting the smelting industry, see John D. Wirth, *Smelter Smoke in North America: The Politics of Transborder*

Pollution (Lawrence: University Press of Kansas, 2000), 3, 45–48, 50–52, 80, 96, 120. For more on ASARCO's support of Swain, see Aiken, "Western Smelters," 502–522.

34. Swain, "Smoke and Fume Investigations"; E. E. Thum, "Smoke Litigation in Salt Lake Valley," *Chemical and Metallurgical Engineering* 22 (1920): 1145–1150; and "Swain's Report on Smoke in Salt Lake Valley," *Chemical and Metallurgical Engineering* 24 (March 16, 1921): 463–465. For more on when Swain sided with the industry, see, for example, Wirth, *Smelter Smoke in North America*.

35. M. Katz and R. J. Cole, "Recovery of Sulfur Compounds from Atmospheric Contaminants," *Industrial and Engineering Chemistry* 42 (1950): 2258–2269. In the 1930s ASARCO built a pilot plant at Garfield that used an ammonia process to remove sulfur dioxide. Corrosion and cost were cited as problems with this technology. In 1927 ASARCO also built a pilot plant at Tacoma that used compression and liquefaction and had a one-ton-per-day capacity. This process was found to be "uneconomic" due to high fuel costs. For a more general discussion of the nonferrous smelting industry's approach to sulfur dioxide control, see R. E. Swain, "Waste Problems in the Nonferrous Smelting Industry," *Industrial and Engineering Chemistry* 31 (November 1939): 1358–1361.

36. M. L. Quinn, "Industry and Environment in the Appalachian Copper Basin, 1890–1930," *Technology and Culture* 34 (1993): 575–612. At Ducktown, recovered sulfuric acid was used in fertilizer manufacture. But western smelter owners resisted this solution, contending that regional differences made sulfuric acid production from sulfur dioxide waste infeasible. In the west, they argued, the distances between smelters and chemical plants were too great, freight costs too high, and a market for fertilizer lacking. Weak streams of sulfur dioxide also presented technical problems. The Trail Smelter in British Columbia was an exception to a general pattern of limited sulfur dioxide control in the west. The Trail Smelter, subject to an international commission's oversight, cut its sulfur dioxide emissions to 350 tons per day, a reduction of two-thirds, by 1937. On resistance to and problems recovering sulfur dioxide see MacMillan, *Smoke Wars*, 170–72; and Swain, "Waste Problems." On Trail Smelter, see Wirth, *Smelter Smoke in North America*, 5.

37. See, for example, "Report of the Selby Smelter Commission," *Journal of Industrial and Engineering Chemistry* 7 (January 1915): 41–45; "Swain's Report on Smoke in Salt Lake Valley"; Wirth, *Smelter Smoke in North America*; and MacMillan, *Smoke Wars*.

38. A. E. Wells, "Results of Recent Investigations of the Smelter Smoke Problem," *Journal of Industrial and Engineering Chemistry* 9 (July 1917): 640–646, at 640.

39. Marcosson, *Metal Magic*, 256–259.

40. R. E. Swain, "Atmospheric Pollution by Industrial Wastes," *Industrial and Engineering Chemistry* 15 (1923): 296–301, at 297. Swain went on to publish further on this topic in 1936, reporting on experiments performed with wheat in his laboratory at Stanford—the results of which argued against the invisible injury theory. ASARCO was acknowledged for providing grant funding for the research. See R. E. Swain and A. B. Johnson, "Effect of Sulfur Dioxide on Wheat Development: Action at Low Concentrations," *Industrial and Engineering Chemistry* 28 (1936): 42–47.

41. Wirth, *Smelter Smoke in North America*.

42. WHO Regional Office for Europe, "Chapter 10: Effects of Sulfur Dioxide on Vegetation: Critical Levels," *Air Quality Guidelines*, 2nd ed. (Copenhagen, Denmark: 2000); and EPA, *Integrated Science Assessment for Oxides of Nitrogen and Sulfur—Ecological Criteria*, EPA/600/R-08/082F (Research Triangle Park, NC: EPA, 2008).

43. The findings from ASARCO's research in Utah are discussed in Wells, "Results of Recent Investigations."

44. Aiken, "Western Smelters," 502–522; and Marcosson, *Metal Magic*, 258–259. Marcosson's work is a company-sponsored history of ASARCO. On the efficacy of high stacks, Marcosson writes: "No disclosure of the ASARCO research program has been more vital to the solution of the smoke problem than this matter of smokestack altitude. It proved to be a major key. Prior to the Company investigation, smelter stacks rarely exceeded 200 feet in height. . . . The combination of high stack and high temperature gives buoyancy to the emerging gases, causing them to rise higher in the air. The answer, therefore, was: 'Build higher stacks.' The result was the 445-foot stack at Murray, the 605-foot stack at Selby, the 573-foot stack at Tacoma, and the 451-foot stack at Federal" (258–259). See also G. R. Hill, M. D. Thomas, and J. N. Abersold, "High Stacks Overcome Concentrations of Gases," *Mining Congress Journal* 31 (1945): 21. This is an article written by ASARCO's research department extolling the benefits of tall stacks.
45. Wirth, *Smelter Smoke in North America*, 5–6.
46. Fulton, *Metallurgical Smoke*, 44–45.
47. Timothy LeCain, "The Limits of 'Eco-Efficiency': Arsenic Pollution and the Cottrell Electrical Precipitator in the U.S. Copper Smelting Industry," *Environmental History* 5 (July 2000): 336–351.
48. Fulton, *Metallurgical Smoke*, 44.
49. Ibid., 32.
50. U.S. EPA, Office of Air Quality Planning and Standards. *Draft EIS: Inorganic Arsenic Emissions from High-Arsenic Primary Copper Smelters-Background Information for Proposed Standards,* April 1983, EPA-450/3-83-009a, Ecology electronic files.
51. See Wells, "Results of Recent Investigations." Wells writes: "The higher the temperature of the gases in the stack, of course, the greater the amount of arsenic and sulfuric acid mist that will escape, and in cases where large quantities of these are present, there is a maximum limit of temperature at which precipitation should be effected. . . . It can be stated that at many lead, copper and zinc smelting plants where a recovery of the dust and fume is desirable, it would be much better for the smoke conditions in the immediate vicinity of the plant if only partial clearness of dust and metallic fumes were obtained, that is to say, 85 or 95 per cent, at a temperature which would still leave the gases much higher in temperature than the atmosphere, rather than to obtain perfect clearness" (645–646). See also Utah Department, General Manager's Office [name illegible] to Mr. Karl Eilers, Vice President, American Smelting and Refining Company, July 27, 1916, *Branin v. ASARCO*.
52. Arthur G. McKee and Company, *Systems Study for Control of Emissions: Primary Nonferrous Smelting Industry*, Vol. 1 (San Francisco: June 1969), prepared for the Division of Process Control, Engineering, NAPCA, Public Health Service, U.S. Department of Health, Education, and Welfare.
53. Advertising and Public Relations Department of ASARCO, *ASARCO News, Tacoma Supplement* (Winter 1969). The newsletter states: "Highly qualified independent consultants have recommended that ASARCO build a new tall chimney 1,000 to 1,200 feet high near the site of the present chimney. Consultants predict impressive reductions in ground-level concentrations of sulfur dioxide. Construction of the chimney is expected to take two years."
54. D. A. Faulkner, "The Effect of a Major Emitter on the Rain Chemistry of Southern British Columbia—A Preliminary Analysis, Part I" (Scientific Services Division, Atmospheric Environment Service, Pacific Region, Environment Canada: April 1, 1987), Ecology electronic files.
55. Glass, *Credible Evidence Report*.

56. C. H. Hine, MD, PhD, Consultant in Occupational Medicine and Industrial Toxicology, to K. W. Nelson, October 25, 1972, *Branin v. ASARCO*. Includes notes of a meeting with Dr. Sherman Pinto, smelter physician.
57. Ibid.
58. Robert Coughlin to Randy Smith, Region X EPA, "Report: ASARCO Closure," January 7, 1985, Ecology electronic files; and Glass, *Credible Evidence Report*.
59. Estimates of historic emissions for the Tacoma Smelter have not been made in the published literature. Gregory Glass, author of the Washington State Department of Ecology's *Credible Evidence Report*, invites readers to make comparisons with Anaconda, which emitted up to seventy-five tons of arsenic per day prior to installing Cottrell electrostatic precipitators.
60. "*Andersen v. Tacoma Smelting Company*, No. 41843," Verdict, December 10, 1917, Superior Court, Pierce County, Washington.
61. Deposition of R. W. Doane, "*Andersen v. Tacoma Smelting Company*, No. 41843," November 28, 1917, Superior Court, Pierce County, Washington.
62. For a graphic description of acute poisoning from arsenic, see James C. Whorton, *The Arsenic Century: How Victorian Britain Was Poisoned at Home, Work, and Play* (New York: Oxford University Press, 2010), 11–16.
63. M. N. Mead, "Arsenic: In Search of an Antidote to a Global Poison," *Environmental Health Perspectives* 113 (June 2005), A378–86.
64. M. Argos, T. Kalra, P. Rathouz, Y. Chen, B. Pierce, F. Parvez, Tariqul Islam, et al., "Arsenic Exposure from Drinking Water, and All-Cause and Chronic-Disease Mortalities in Bangladesh (HEALS): A Prospective Cohort Study," *Lancet* 376 (July 2010): 252–258.
65. Whorton, *Arsenic Century*.
66. M. O. Varner, Director, Department of Environmental Sciences, ASARCO, to Dr. Samuel Milham, December 12, 1983, Ecology electronic files; and M. O. Varner to Dr. Samuel Milham, January 23, 1984, Ecology electronic files.
67. Bostwick R. Ketchum, Woods Hole Oceanographic Institute, *Evaluation of the Proposal to Dispose of Arsenic Waste by Discharge at Sea*, March 1, 1953, *Branin v. ASARCO*.
68. D. R. Girdwood, President, Girdwood Shipping Company, to Mr. E. F. Eldridge, Director & Chief Engineer, State of Washington Pollution Control Commission, April 10, 1953, Washington State Archives, hereafter WSA, ASARCO Tacoma Industries, 1949–56, Box 13, Water Pollution Control Commission, 71-5-376.
69. Whorton, *Arsenic Century*; and W. C. Hueper, *Occupational Tumors and Allied Diseases* (Springfield, IL: Charles C. Thomas, 1942).
70. Whorton, *Arsenic Century*.
71. James C. Whorton, *Before Silent Spring: Pesticides and Public Health in Pre-DDT America* (Princeton, NJ: Princeton University Press, 1974), esp. 177–181.
72. Whorton, *Before Silent Spring*; and F. Roth, "The Sequelae of Chronic Arsenic Poisoning in Moselle Vintners," *German Medical Monthly* 2 (1957): 172–175.
73. Whorton, *Before Silent Spring*, 21.
74. Hueper, *Occupational Tumors and Allied Diseases*.
75. Rachel Carson, *Silent Spring* (New York: Houghton Mifflin Company, 1962).
76. C. Potera, "Food Safety: U.S. Rice Serves up Arsenic," *Environmental Health Perspectives* 11 (2007): A296.
77. Marcosson, *Metal Magic*, 114.
78. H. Y. Walker, Tacoma Smelter Manager, to Mr. Karl Eilers, Vice President, ASARCO, September 20, 1916, *Branin v. ASARCO*.

79. Utah Department, General Manager's Office [name illegible] to Mr. Karl Eilers, Vice President, American Smelting and Refining Company, July 27, 1916, *Branin v. ASARCO.*
80. Ibid.
81. "New York Concern to Build Chimney: Work Started Soon on Tall Stack," *Tacoma News Ledger*, February 4, 1917, 24; and E. A. Peters, "Smelter Spends $5,000 to Bring Novel World Record to Tacoma," *Tacoma Times*, October 13, 1917, 3.
82. "More Complaints of Smoke Nuisance," *Tacoma Daily Ledger*, August 3, 1917, 3.
83. National Air Pollution Control Techniques Advisory Committee, U.S. EPA, Office of Air, Noise and Radiation Emission Standards and Engineering Division, "Minutes of Meeting March 17 and 18, 1981," April 17, 1981, WSA, Ecology, Box 372, F: Correspondence 1978–1981; and R. G. Tyler, *Report on a Smoke Abatement Investigation for Tacoma, Washington* (1948), available at the Tacoma Public Library.
84. Whorton, *Before Silent Spring*; and MacMillan, *Smoke Wars*, 243–244.
85. P. M. Tyler and A. V. Petar, *Arsenic, Bureau of Mines, Economic Paper 17* (U.S. Department of the Interior, Washington, DC: Government Printing Office, 1934), 31. Explaining how the smelting industry makes decisions regarding arsenic capture Tyler and Petar write: "Inasmuch as the production of arsenical dusts at copper and lead smelters has generally exceeded the demand for arsenic that might be recovered therefrom, upward movements in prices are normally met promptly by increased production. On the other hand, a lower limit is placed upon prices at the point where the producers find it unprofitable to work up their dusts into saleable production. Stocks of refined arsenic rarely reach unwieldy proportions, and as most of the producers are largely well financed concerns, there is no inducement for them to dump their product on the market or take a heavy loss on sales."
86. Unger and Unger, *Guggenheims*, 133–136.
87. "Helps Alaska Copper Men," *New York Times*, September 22, 1911, 7; "American Smelting Plants Rushed to the Limit," *Wall Street Journal*, August 19, 1915, 5; and "American Smelting," *Wall Street Journal*, March 28, 1917, 1.
88. "Smelter Fume Parley Held: No Relief Can Be Expected, City Officials Report, after Conference," *Tacoma Daily Ledger*, August 5, 1927, 1.
89. See, for example, Armand Labbe, "Presentation before the Board of Directors of the Puget Sound Air Pollution Control Agency at the Public Hearing for Proposed Amendments to Regulation I for Controlling the Emission of Arsenic," February 14, 1973, ASDSF 13.5; Herbert G. Lawson, "Dilemma in Tacoma: Smelter's Emissions Threaten Populace, But So Does Possible Loss of 1,000 Jobs," *Wall Street Journal*, July 16, 1975, 36; "Smelter Promises to Keep Trying," *Vashon Maury Island Beachcomber*, January 22, 1976; and "Asarco Awaits Decision on Variance for Operation of Tacoma Copper Smelter," *Metals Week*, November 9, 1981.
90. Aiken, "Western Smelters," 502–522.
91. "Report of the Selby Smelter Commission."
92. "Smelter Fume Parley Held," 1.
93. Tyler and Petar, *Arsenic, Bureau of Mines, Economic Paper 17.*
94. L. V. Olson, Director Department of Agricultural Research, to D. A. Somerville, Tacoma Smelter, February 7, 1951, *Branin v ASARCO.*
95. "Ruston: City Has Low Tax Rate," *Tacoma Sunday Ledger-News Tribune*, May 25, 1947, 4.
96. Tyler, *Report on a Smoke Abatement Investigation*, 25.
97. D. A. Somerville to Mr. L. V. Olson, "Memorandum: Franklin E. Kinsey," January 26, 1954, *Branin v. ASARCO.*

98. D. A. Somerville to Mr. L. V. Olson, September 11, 1953, *Branin v. ASARCO*.
99. C. H. Hine, MD, PhD, Consultant in Occupational Medicine and Industrial Toxicology, to K. W. Nelson, October 25, 1972, *Branin v. ASARCO*. Includes notes of a meeting with Dr. Sherman Pinto, smelter physician.
100. M. O. Varner, Director, Department of Environmental Sciences, ASARCO, to Dr. Samuel Milham, December 12, 1983, Ecology electronic files.
101. In reference to authoritative sources, see Wells, "Results of Recent Investigations of the Smelter Smoke Problem," 640–646; and R. E. Swain, "Atmospheric Pollution by Industrial Wastes," 296–301; on "limited public information about their emissions," see L. V. Olson, Director Department of Agricultural Research, to D. A. Somerville, Tacoma Smelter, February 7, 1951, *Branin v ASARCO*.
102. See, for example, "Swain's Report on Smoke in Salt Lake Valley," *Chemical and Metallurgical Engineering* 24 (March 16, 1921): 463–465; "Smelter Fume Parley Held," 1; and L. V. Olson, Director Department of Agricultural Research, to D. A. Somerville, Tacoma Smelter, February 7, 1951, *Branin v ASARCO*.

CHAPTER 2 CITY OF DESTINY, CITY OF SMOKE

1. Alfred Cavanagh, "A City of Industry: Tacoma's Tidelands Promise Her the Manufacturing Supremacy of the Pacific Coast," *Harper's Weekly* 57 (May 31, 1913): 13.
2. Murray Morgan, *Puget's Sound: A Narrative of Early Tacoma and the Southern Sound* (Seattle: University of Washington Press, 1979).
3. Cavanagh, "City of Industry," 13.
4. R. G. Tyler, *Report on a Smoke Abatement Investigation for Tacoma, Washington* (1948), accessed at Tacoma Public Library.
5. E. R. Hendrickson, D. M. Keagy, and R. L. Stockman, *Evaluation of Air Pollution in the State of Washington: Report of Cooperative Survey Made July 1 through November 30, 1956* (U.S. Department of Health Education and Welfare Public Health Service and State of Washington Department of Health, December 1956).
6. Tyler, *Report on a Smoke Abatement Investigation*.
7. Timothy Egan, "Tacoma Journal: On Good Days, the Smell Can Hardly Be Noticed," *New York Times*, April 6, 1988, A16.
8. Morgan, *Puget's Sound*, 328.
9. "Tideflat Odor Is Under Fire: Council Takes First Steps to End Tacoma's Municipal Halitosis," *Tacoma Daily Ledger*, March 23, 1930, 1.
10. Morgan, *Puget's Sound*, 329.
11. David Stradling, *Smokestacks and Progressives: Environmentalists, Engineers, and Air Quality in America, 1881–1951* (Baltimore: Johns Hopkins University Press, 1999), 159.
12. "No Cure of Pulp Odor: Union Plant Manager Says Everything Possible Being Done Here," *Tacoma News Tribune*, April 15, 1930, A3.
13. Scott H. Dewey, *Don't Breathe the Air: Air Pollution and U.S. Environmental Politics, 1945–1970* (College Station: Texas A&M University Press, 2000), 8–9.
14. Stradling, *Smokestacks and Progressives*, 61–84, 182–191.
15. Carlos A. Schwantes, *The Pacific Northwest: An Interpretive History* (Lincoln: University of Nebraska Press, 1989), 408–415.
16. U.S. Bureau of the Census, Table 17, "Population of the 100 Largest Urban Places: 1940," Internet release date June 15, 1998, http://www.census.gov/population/www/documentation/twps0027/tab17.txt.

17. Casey Davisson, "Melting Pot for the World—That's Tacoma's Smelter: Seven Days a Week for Half Century," *Tacoma Times*, February 27, 1940, 11.
18. Tyler, *Report on a Smoke Abatement Investigation*.
19. "Fumes Are Protested: Garden Club Says Smelter Killing Gardens, Lawns," *Tacoma News Tribune*, August 19, 1942, 1.
20. "Tacoman Plans Suit over Smelter's Fumes," *Tacoma News Tribune*, August 3, 1945, 1.
21. "Reported Here a Year Ago," *Tacoma News Tribune*, April 27, 1945, 3; A. H. Newman, "Fumes," *Tacoma News Tribune*, July 30, 1945, 4; and Paul O. Anderson, "Warns of Air Pollution," *Tacoma News Tribune*, November 1, 1948, 6.
22. Samuel P. Hays, *Beauty Health and Permanence: Environmental Politics in the United States, 1955-1985* (Cambridge: Cambridge University Press, 1987), 2–5; and Samuel P. Hays, *Explorations in Environmental History* (Pittsburgh: University of Pittsburgh Press, 1998), 336–351.
23. B. Nemery, P. H. Hoet, and A. Nemmar, "The Meuse Valley Fog of 1930: An Air Pollution Disaster," *Lancet* 357 (March 3, 2001): 704–708; L. P. Snyder, "The Death-Dealing Smog over Donora, Pennsylvania: Industrial Air Pollution, Public Health Policy, and the Politics of Expertise, 1948–1949," *Environmental History Review*, Spring 1994, 117–139; and M. L. Bell and D. L. Davis, "Reassessment of the Lethal London Fog of 1952: Novel Indicators of Acute and Chronic Consequences of Acute Exposure to Air Pollution," *Environmental Health Perspectives* 109, Suppl. 3 (June 2001): 389–394.
24. On the Donora event and funding for air pollution research, see Snyder, "Death-Dealing Smog." On the investigations see, for example, Clarence A. Mills, *This Air We Breathe* (Boston: Christopher Publishing House, 1962); L. Breslow and J. Goldsmith, "Health Effects of Air Pollution," *American Journal of Public Health* 48 (1958): 913–917; W. C. Hueper, "Environmental Causes of Lung Cancer," *Public Health Reports* 71 (1956): 94–98. On the popular media's coverage of the events, see Dewey, *Don't Breathe the Air*, 136–137, 248–249.
25. Dewey, *Don't Breathe the Air*, 248–249.
26. See ibid., 226–254, for a discussion of the role of the federal government in air pollution control in the 1950s and 1960s.
27. "Mayor Hears Plan to Curb Smoke Here," *Suburban Times*, December 23, 1947, 1.
28. "City to Make First Move in Ban on Smoke," *Tacoma Ledger*, December 7, 1947, A23.
29. Tyler, *Report on a Smoke Abatement Investigation*, 2.
30. Ibid., 2–3.
31. Mrs. R. L. Doud to Tacoma Chamber of Commerce, July 28, 1947, Washington State Historical Society, hereafter WSHS, Tacoma Chamber of Commerce Manuscript Collection, Ms40.
32. Marshall Ramstad, Chemical Engineer, Tacoma Chamber of Commerce, to E. R. Marble, Manager, Tacoma Smelter, November 26, 1947, WSHS, Tacoma Chamber of Commerce Manuscript Collection, Ms40; E. R. Marble, Manager, Tacoma Smelter, to Marshall Ramstad, Chemical Engineer, Tacoma Chamber of Commerce, December 29, 1947, WSHS, Tacoma Chamber of Commerce Manuscript Collection, Ms40; F. J. Walsh, Industrial Engineer, Tacoma Chamber of Commerce, to M. B. Littlefield, Industrial Development Department, ASARCO, July 14, 1947, WSHS, Tacoma Chamber of Commerce Manuscript Collection, Ms40; and M. B. Littlefield, Industrial Development Department to F. J. Walsh, Industrial Engineer, Tacoma Chamber of Commerce, July 21, 1947, WSHS, Tacoma Chamber of Commerce Manuscript Collection, Ms40.

33. Puget Sound Air Pollution Control Agency, *Air Contaminant Emissions from ASARCO's Tacoma Smelter, Chronology of Control Efforts* (n.d.), obtained from Puget Sound Clean Air Agency.
34. "New Smoke Measure Offered: Anti-Air Pollution Ordinance Provides for Regulation of Smoke, Dust, Fumes," *Tacoma News Tribune*, October 24, 1949, 1.
35. Ibid.; and John Rosene, Division of Air Pollution Control, *Special Municipal Report: Significance of Air Pollution Control Tacoma, Washington 1950–1953* (1953), 5, available at the Tacoma Public Library.
36. "Law to Aid Smoke Ban," *Tacoma News Tribune*, January 15, 1950, 1.
37. E. F. Eldridge, Director and Chief Engineer, State of Washington Pollution Control Commission, to Dr. Lewis J. Cralley, Senior Scientist, Industrial Hygiene Field Headquarters, U.S. Public Health Service, September 11, 1950, Washington State Archives, hereafter WSA, Department of Health, Director's Office, RS9 Subject Files 1936–1956 Box 79, F: Pollution Commission 1949–1950.
38. "Air Pollution Law Held Up," *Tacoma News Tribune*, February 18, 1951, A11.
39. Manufacturing Chemists Association, "Agenda: Executive Committee Meeting," September 9, 1947, *Chemical Industry Archives*, http://www.chemicalindustryarchives.org/search/pdfs/cma/19470909_00000101.pdf.See also, for example, Manufacturing Chemists Association, "Minutes of Meeting: Air Pollution Abatement Committee," December 7, 1950, p. 1–5; and March 14, 1951, p. 11–14, *Chemistry Industry Archives*, http://www.chemicalindustryarchives.org/search/pdfs/cma/19501207_00001614.pdf.
40. "Industry Advised on Air Pollution: Manufacturing Chemists Say That Waste Alone Is Not to Blame for Fouling," *New York Times*, February 27, 1952, 27.
41. See, for example, "ASARCO Fears Shutdown," *Tacoma News Tribune*, January 27, 1976, A1; Margaret Ainscough, "Asarco Letter to Employees Blasted," *Tacoma News Tribune*, April 14, 1977, B1; Roger Ainsley, "How Safe Should the Smelter Be?" *Seattle Post Intelligencer, Northwest Magazine*, March 9, 1980, 5; Bob Lane, "Smelter Firm Again Threatens Shutdown," *Seattle Times*, May 4, 1971, C15; Jack Pyle, "Smelter Needs Aid, Says Federal Official," *Tacoma News Tribune*, April 8, 1977, A1; and "Tacoma Smelter Threatens to Close," *Seattle Post-Intelligencer*, August 6, 1983, C1.
42. "Smoke Control in Doldrums," *Tacoma Sunday News Tribune and Ledger*, October 11, 1953, A12.
43. George R. Hill, "Report: Condition of Vegetation at Tacoma, August 26–27, 1952," *Branin v. ASARCO*, 93-CV-5132, Accession # 021-04-0102, Box 1-9, Location # 3050865, settlement papers obtained at Seattle Federal Records Center; hereafter referred to as *Branin v. ASARCO*.
44. "Our Own Brand of Fog," *Tacoma News Tribune*, November 29, 1953, B4.
45. J. L. Jones, MD, Chief, Division of Preventive Medical Services, to O. C. Hopkins, Sanitary Engineer Director, Department of Health, Education and Welfare, Public Health Service, July 15, 1954, WSA, DOH Director's Office, RS9 Subject Files, 1936–56, Air Pollution, Box 36, F: Air Pollution Control 1948–60.
46. Robert L. Stockman, to Files, "Memorandum: Meeting Re Air Pollution Control Legislation (February 1, 1947)," February 4, 1957, WSA, DOH Director's Office RS10 Subject Files 1955–62 Box 114 F: Air Pollution; and "Summary of Recent Conference to Consider Development of Industrial Health Program for the State of Washington" (meeting held September 16, 1954), WSA, DOH Director's office, RS4 Organizations and Associations, Box 17, F: Industrial Health Program State of Washington.
47. "Tacoma Aroma to Stay around for Awhile Yet," *Tacoma News Tribune*, October 21, 1956, 1.

48. Tyler, *Report on a Smoke Abatement Investigation*, 9–10; and R. J. Vong, T. V. Larson, and W. H. Zoller, "A Multivariate Chemical Classification of Rainwater Samples," *Chemometrics and Intelligent Laboratory Systems* 3 (1988): 99–109.
49. Hendrickson, Keagy, and Stockman, *Evaluation of Air Pollution*; and A. R. Dammkoehler, *History of Relations: Puget Sound Air Pollution Control Agency and American Smelting and Refining Company Copper Smelter March, 1968–October, 1970* (Report issued by Puget Sound Air Pollution Control Agency, October 20, 1970).
50. "Tacoma Aroma to Stay around for Awhile Yet," 1.
51. U.S. Bureau of the Census, Table 17, 1940; and U.S. Bureau of the Census, Table 19, "Population of the 100 Largest Urban Places: 1960," Internet release date June 15, 1998, http://www.census.gov/population/www/documentation/twps0027/tab19.txt.
52. D. A. Somerville to L. V. Olson, January 31, 1956, *Branin v. ASARCO*.
53. R. E. Shinkoskey to L. V. Olson, December 11, 1957, *Branin v. ASARCO*.
54. Charles Wolverton, "New Storm Drains Would Permit Big Development," *Tacoma News Tribune*, January 29, 1956, A8; and Dean St. Dennis, "Hundreds of Homes to Be Built in West End," *Tacoma Sunday News Tribune and Ledger*, March 11, 1956, C14.
55. Interview with Judy Alsos, June 15, 2006, Tacoma, Washington. Digital audio recording. File in author's possession.
56. Hilman Ratsch, *Heavy-Metal Accumulation in Soil and Vegetation from Smelter Emissions*, August, 1974, U.S. EPA National Environmental Research Center, Office of Research and Development, EPA-660/3-74-012, 2.
57. Interview with Judy Alsos, June 15, 2006, Tacoma, Washington.
58. U.S. EPA, *Integrated Science Assessment for Sulfur Oxides—Health Criteria*, Final Report, September 2008. Washington, D.C., EPA/600/R-08/047F; and American Academy of Pediatrics, Committee on Environmental Health, "Ambient Air Pollution: Health Hazards to Children," *Pediatrics* 114 (2004): 1699–1707.
59. On Drinker's role as director of ASARCO's Department of Hygiene, see Michael O. Varner, *ASARCO Incorporated: The Department of Environmental Sciences* (n.d.), *Branin v. ASARCO*. On the conflict over Amdur's sulfur dioxide research, see Devra Davis, *When Smoke Ran Like Water: Tales of Environmental Deception and the Battle against Pollution* (New York: Basic Books, 2002), 66–77, quotes at 74 and 75.
60. Interview with Judy Alsos, June 15, 2006, Tacoma, Washington.
61. Robert Dales, Richard T. Burnett, Marc Smith-Doiron, and David M. Stieb, "Air Pollution and Sudden Infant Death Syndrome," *Pediatrics* 113 (2004): e628–e631.
62. Hill, "Report: Condition of Vegetation"; and State Air Pollution Control Board, "Minutes of Meeting," December 27, 1961, WSA, DOH Director's Office, RS10 Subject Files, 1955–62, Box 114, F: APCB, 1961–63.
63. Judy Alsos to Bernard Bucove, July 19, 1961, WSA, DOH Director's Office, RS10 Subject Files, 1955–1962, Box 114 F: Air Pollution, 1962.
64. "Primary National Ambient Air Quality, Standard for Sulfur Dioxide," 75 Fed. Reg. 119 (June 22, 2010), 35520. The new one-hour peak standard is 75 ppb. Peak sulfur dioxide concentrations measured in Tacoma in the 1950s were between 3 and 5 ppm (3,000–5,000 ppb).
65. L. White, "That Creeping Menace Called Smog," *Saturday Evening Post*, May 4, 1957, 122.
66. L. V. Olson to R. D. Bradford, "Confidential Memorandum: Tacoma Smoke," June 6, 1957, *Branin v. ASARCO*; and D. A. Somerville to R. E. Shinkoskey, "Personal and Confidential Memorandum: Smoke," June 4, 1957, *Branin v. ASARCO*.
67. Olson to Bradford, "Confidential Memorandum."

68. Bernard Bucove, MD, Director, DOH to DHEW, Public Health Service, Community Air Pollution Program, September 23, 1957, WSA, DOH Director's Office, RS10 Subject Files, 1955–1962 Box 114 F: Air Pollution.
69. Interview with Judy Alsos, June 15, 2006, Tacoma, Washington.
70. D. A. Somerville to L. V. Olson, February 26, 1960, *Branin v. ASARCO*. Survey attached.
71. E. McL. Tittman, Executive Vice President, to R. E. Shinkoskey, Tacoma Smelter Manager, February 29, 1960, *Branin v. ASARCO*.
72. R. E. Shinkoskey to E. McL. Tittman, March 3, 1960, *Branin v. ASARCO*.
73. "Industrial Firms Oppose State Air Pollution Bill," *Tacoma News Tribune*, March 12, 1960, 3.
74. "Tacoma Smells Better than It Used To, Or Else We've Gotten Used to Tacoma's Smell," *Tacoma News Tribune*, November 20 1960, 1.
75. State Air Pollution Control Board, "Minutes of Meeting," December 27, 1961, WSA, DOH Director's Office, RS10 Subject Files, 1955–62 Box 114, F: APCB, 1961–63, statement of Judy Alsos to the governor, December 4, 1961, attached.
76. Interview with Judy Alsos, June 15, 2006, Tacoma, Washington.
77. State Air Pollution Control Board, "Minutes of Meeting," December 27, 1961.
78. Robert L. Stockman to Mrs. Alsos, July 6, 1961, WSA, DOH Director's Office, RS10 Subject Files, 1955–62. Box 114, F: Air Pollution, 1962.
79. State Air Pollution Control Board, "Minutes of Meeting," December 27, 1961.
80. "Pollution Talk Is County Aim," *Tacoma News Tribune*, July 25, 1961, A11.
81. Robert L. Stockman to Robert Shinkoskey, August 18, 1961, WSA, DOH Director's Office, RS10 Subject Files 1955–62 Box 114 F: Air Pollution 1962.
82. Charles Barber, President, ASARCO, to Richard Falknor, Office of Senator Henry M. Jackson, U.S. Senate, November 24, 1961, WSA, DOH Director's Office, RS 10 Subject Files 1955–1962 Box 114, F: APCB 1961–63.
83. Bernard Bucove, MD, to Gov. Albert D. Rossellini, "Memorandum: Your Appointment with Mrs. Judy Alsos," December 1, 1961, WSA, DOH Director's Office, RS10 Subject Files, 1955–62 Box 114, F: Air Pollution, 1962.
84. Denny MacGougan, "City Sets Sights on Smelter," *Tacoma News Tribune*, May 3, 1967, 1.
85. Kay MacDonald, governor's secretary, to Judy Alsos, July 28, 1961, WSA, DOH Director's Office, RS10 Subject Files, 1955–62, Box 114, F: Air Pollution, 1962.
86. "State Urged to Scan Smelter Air Pollution," *Tacoma News Tribune*, August 30, 1961, A2.
87. Bucove to Rossellini, "Memorandum: Your Appointment with Mrs. Judy Alsos."
88. "Smelter Fallout Protest Filed," *Tacoma News Tribune*, December 5, 1961, 31; and interview with Judy Alsos, June 15, 2006, Tacoma, Washington.
89. State Air Pollution Control Board, "Minutes of Meeting," December 27, 1961, statement of Judy Alsos to the governor, December 4, 1961, attached.
90. Ibid.
91. "Smelter Fallout Protest Filed," 31; and State Air Pollution Control Board, "Minutes of Meeting," December 27, 1961.
92. Bernard Bucove to Judy Alsos, December 22, 1961, WSA, DOH, Director's Office RS10 Subject Files 1955–62 Box 114, F: Air Pollution, 1962.
93. State Air Pollution Control Board, "Minutes of Meeting," December 27, 1961; and Robert L. Stockman, State DOH to C. R. Fargher, MD, MPH Director, Tacoma-Pierce County Health Department, March 4, 1960, WSA, DOH Director's Office, RS10 Subject Files 1955–62. Box 114, F: Air Pollution, 1962.
94. Barber to Falknor, November 24, 1961, WSA.

95. R. E. Shinkoskey to Albert D. Rossellini, December 18, 1961, WSA, DOH, Director's Office, RS10 Subject Files, 1955–62, Box 114, F: APCB, 1961–63.
96. R. E. Shinkoskey to W. G. Rouillard, General Manager, Salt Lake City Office, "Memorandum: Control of Atmospheric Pollutants," June 6, 1961, *Branin v. ASARCO*.
97. L. V. Olson to D. A. Somerville, October 4, 1961, *Branin v. ASARCO*.
98. S. M. Lane, Manager, East Helena Plant, to W. G. Rouillard, "Memorandum: Horses—East Helena," August 11, 1965, *Branin v. ASARCO*; and "Handwritten Notes: WJB Smith—Dogs," April 11–April 15, 1969, *Branin v. ASARCO*.
99. David D. Rowlands to Luther Terry, Surgeon General, U.S. Public Health Service, February 16, 1962, WSA, DOH Director's Office, RS10 Subject Files, 1955–1962, Box 114, F: Air Pollution, 1962.
100. Ibid.
101. "U.S. Enters Air Pollution Issue," *Suburban Times*, May 9, 1962, 1.
102. "U.S. Experts Open Pollution Probe Here," *Tacoma News Tribune*, May 1, 1962, 9.
103. Jean J. Schueneman, Chief, Technical Assistance Branch, Division of Air Pollution, U.S. Public Health Service, to David Rowlands, June 7, 1962, WSA, DOH Director's Office, RS10 Subject Files, 1955–1962, Box 114, F: Air Pollution, 1962.
104. See Charles O. Jones, *Clean Air: the Policies and Politics of Pollution Control* (Pittsburgh: University of Pittsburgh Press, 1975), 113. Jones notes that in 1961 "only seventeen states and eighty-five local agencies were spending as much as $5,000 on air pollution programs." Almost half of all local and state personnel in air pollution control were employed in California.
105. Dewey, *Don't Breathe the Air*, 226–254.
106. L. V. Olson to S. M. Lane, October 14, 1965, *Branin v. ASARCO*.
107. Philip Shabecoff, *A Fierce Green Fire: The American Environmental Movement* (Washington, DC: Island Press, 2003), 103–119.
108. "Weather Better, Pollution Worse," *Tacoma News Tribune*, August 1966.
109. R. E. Shinkoskey to K. W. Nelson, Director Hygiene and Agricultural Research, July 28, 1966, *Branin v. ASARCO*.
110. Denny MacGougan, "Industrial People Mostly Mum at Meeting with City Council on Air Pollution Problems," *Tacoma News Tribune*, October 15, 1966, 1; and Denny MacGougan, "Pollution Well Aired by City Councilmen," *Tacoma News Tribune*, September 28, 1966, 1.
111. Denny MacGougan, "Foes of Air Pollution Storm Budget Talks," *Tacoma News Tribune*, October 4, 1966, 1; and MacGougan, "Industrial People Mostly Mum."
112. "Air Pollution Control Hot Potatoes on Ice," *Tacoma News Tribune*, October 19, 1966, B9; and Washington State Clean Air Act, Chapter 70.94 RCW, http://apps.leg.wa.gov/RCW/default.aspx?cite=70.94.
113. "City Council Passes Ordinance Severely Limiting Smog Level," *Tacoma News Tribune*, July 6, 1967, 1.
114. "New Pollution Act Would Aim At Smelter," *Tacoma News Tribune*, May 24, 1967, 1.
115. Ibid.
116. "Air Effects Overstated, Smelter Says," *Tacoma News Tribune*, May 25, 1967, C17.
117. Ibid.
118. On the smelter's capacity to recover toxic metals, see Puget Sound Air Pollution Control Agency, "Staff Report: Section 9.19 Regulation I Arsenic Emission Standard," February 14, 1973, Environmental Protection Agency Site File, Commencement Bay/Nearshore Tideflats ASARCO Smelter Facility Site File, ASDSF 1.1.1. For a more general

discussion of pollution, see Jack Ryan, "Smelter Stack Belching Pollution," *Seattle Post-Intelligencer*, August 13, 1969, 4.

119. See, for example, Olson to Bradford, "Confidential Memorandum: Tacoma Smoke," June 6, 1957, *Branin v. ASARCO*; "Smelter Officials Tell of Anti-Pollution Steps," *Tacoma News Tribune*, July 27, 1961; Shinkoskey to Rosellini, "Letter," December 18, 1961; State Air Pollution Control Board, "Minutes of Meeting," December 27, 1961; and "U.S. Experts Open Pollution Probe Here."

CHAPTER 3 UNCOVERING A CRISIS IN EL PASO

1. John J. Sheehy, Attorney Rogers & Wells, to Files, "Memorandum: Gulf Resources & Chemical Corporation—ILZRO," October 2, 1974, *Edna Grace Yoss, Raymond Hans Yoss, Arlene Mae Yoss, Richard A. McCartney, Christina M. McCartney, Paula A. McCartney, Raymond E. Dennis and Harley Dennis, by their guardian ad litem, Brian J. Linn, Plaintiffs, v. Bunker Hill Company, a Delaware corporation; Gulf Resources & Chemical Corporation, a Delaware corporation; Defendants*, No. CIV-77–2030, United States District Court for the District of Idaho, hereafter, *Yoss et al. v. Bunker Hill and Gulf*.
2. Philip Shabecoff, *A Fierce Green Fire: the American Environmental Movement* (Washington, DC: Island Press, 2003), 121–123.
3. General Accounting Office, Report to the Chairman, Subcommittee on Oversight and Investigations, Committee on Energy and Commerce, *Air Pollution: Sulfur Dioxide Emissions from Nonferrous Smelters Have Been Reduced*, GAO/RCED-86–91 (Washington, DC, April 1986).
4. Arthur G. McKee and Company, *Systems Study for Control of Emissions, Primary Nonferrous Smelting Industry*, vol. 1. Prepared for Division of Process Control, Engineering, NAPCA, PHS, U.S. Dept. of Health and Human Services (Washington, DC, June 1969).
5. Herbert G. Lawson, "Getting Sick of Sulphur in Tacoma," *Wall Street Journal*, October 14, 1970, 22; and E. W. Kenworthy, "U.S. Agency in Middle of Pollution Dispute in Smelting Industry," *New York Times*, February 7, 1972, 24.
6. Kenworthy, "U.S. Agency in Middle."
7. "Human Lead Absorption—Texas," *Morbidity and Mortality Weekly Report*, December 8, 1973, 405–407, http://www.cdc.gov/mmwr/preview/mmwrhtml/lmrk095.htm.
8. Lawson, "Getting Sick of Sulphur in Tacoma."
9. Sito Negron, "Asarco Announcement: El Paso Smelter Will Not Reopen," *Newspaper Tree*, February 3, 2009, http://www.newspapertree.com/news/3415-asarco-announcement-el-paso-smelter-will-not-reopen, accessed December 21, 2009.
10. Michael E. Ketterer, *The ASARCO El Paso Smelter: A Source of Local Contamination of Soils in El Paso (Texas), Ciudad Juarez (Chihuahua, Mexico), and Anapra (New Mexico)*, Summary Report, January 27, 2006, http://texas.sierraclub.org/air/Sierra%20Club%20ASARCO%20Study.pdf, accessed January 24, 2011.
11. Isaac F. Marcosson, *Metal Magic: The Story of the American Smelting and Refining Company* (New York: Farrar, Straus and Company, 1949), 191; Ketterer, *ASARCO El Paso Smelter*; John W. Drexler, *A Study on the Source of Anomalous Lead and Arsenic Concentrations in Soils from the El Paso Community—El Paso, Texas*, June 5, 2003, http://www.epa.gov/region6/6sf/texas/el_paso/tx_el_paso_finalreport.pdf, accessed January 24, 2011; and C. Olmedo, D. L. Soden, O. Morera, M. McElroy, D. A. Schauer, S. Pena, C. Fuentes, et al., *Valuing the Paso Del Norte: Resident and Business Perspectives on the Value of*

the *Environment Relative to Reopening the ASARCO Copper Smelter*, Technical Report No. 2007-08 (University of Texas at El Paso, January 2008).
12. See Gerald Markowitz and David Rosner, *Deceit and Denial: The Deadly Politics of Industrial Pollution* (Berkeley: University of California Press, 2002), 45–54.
13. This information comes from an untitled, anonymously authored document that provides an overview of the Lead Industries Association and its main activities. Obtained from Professor David Rosner, Columbia University.
14. Jerome F. Cole, Vice President ILZRO, "Deposition," March 7, 1980, 6, 59, *Yoss et al. v. Bunker Hill and Gulf*, Files of Paul Whelan, Plaintiff's attorney.
15. See, for example, James D. Callaghan, Hill & Knowlton, Inc., to Philip E. Robinson, Lead Industries Association, Inc., "Memorandum: Further Distribution of the McNeil Study," July 1, 1974, *Yoss et al. v. Bunker Hill and Gulf*, Files of Paul Whelan, Plaintiff's attorney; James D. Callaghan, Hill & Knowlton, Inc., to Philip E. Robinson, Lead Industries Association, Inc., "Memorandum: Idaho Lead Smelter Story," September 6, 1974, *Yoss et al. v. Bunker Hill and Gulf*, Files of Paul Whelan, Plaintiff's attorney; and ILZRO, "Comments on Testimony at the Hearing on Reconsideration of the California Ambient Air Quality Standard for Lead, California Air Resources Board, Sacramento Nov. 3 & 4, 1975," November 21, 1975, *Yoss et al. v. Bunker Hill and Gulf*, Files of Paul Whelan, Plaintiff's attorney.
16. "Human Lead Absorption—Texas," *MMWR*, December 8, 1973, 405–407.
17. See for example, ILZRO, "Comments on Testimony," November 21, 1975, *Yoss et al. v. Bunker Hill and Gulf*, Files of Paul Whelan, Plaintiff's attorney.
18. See Markowitz and Rosner, *Deceit and Denial*.
19. Committee on Biologic Effects of Atmospheric Pollutants, *Lead: Airborne Lead in Perspective* (Washington, DC: National Academy of Sciences, 1972), 13. In 1968 automobile exhaust put 181,000 tons of lead into the atmosphere.
20. Markowitz and Rosner, *Deceit and Denial*; Christian Warren, *Brush with Death: A Social History of Lead Poisoning* (Baltimore: Johns Hopkins University Press, 2000); and Committee on Biologic Effects, *Lead*, 13.
21. Markowitz and Rosner, *Deceit and Denial*. See, in particular, "Old Poisons, New Problems," 108–138.
22. Ibid.
23. Ibid.
24. Warren, *Brush with Death*, 199–200; and J. L. Steinfeld, "The Surgeon General's Policy Statement on Medical Aspects of Childhood Lead Poisoning," U.S. Department of Health and Human Services, August 1971.
25. Committee on Biologic Effects, *Lead*, 132–139; and Office of Research and Development, "Air Quality Criteria for Lead 1977" (Washington, DC: Environmental Protection Agency, December 1977), 12-5–12-6.
26. Jack Lewis, "Lead Poisoning: A Historical Perspective," *EPA Journal*, May 1985, http://www2.epa.gov/aboutepa/lead-poisoning-historical-perspective.
27. Warren, *Brush with Death*, 220–221.
28. Markowitz and Rosner, *Deceit and Denial*, 109.
29. Telephone interview with Philip Landrigan, January 6, 2006. Notes in author's possession.
30. See S. P. Hays, "The Role of Values in Science and Policy: The Case of Lead," in *Explorations in Environmental History* (Pittsburgh: University of Pittsburgh Press, 1998), 291–311, at 309.

31. Committee on Biologic Effects, *Lead*, 132, 181.
32. Wolfgang Saxon, "B. Carnow, 74, A Specialist in Health Issues," *New York Times*, November 10, 1996, http://www.nytimes.com/1996/11/10us/b-carnow-74-a-specialist-in-health-issues.html?pagewanted=1, accessed December 21, 2009.
33. Office of Research and Development, *Air Quality Criteria for Lead 1977*, appendix E, "Abstract of a Review of Three Studies on the Effects of Lead Smelter Emissions in El Paso, Texas."
34. Stephen Thacker, Donna Stroup, and David Sencer, "Epidemic Assistance by the Centers for Disease Control and Prevention: Role of the Epidemic Intelligence Service, 1946–2005," *American Journal of Epidemiology* 174, suppl. 11 (2011): S4–S15.
35. John J. Sheehy to Files, "Memorandum: Gulf Resources & Chemical Corporation—ILZRO," October 2, 1974, *Yoss et al. v. Bunker Hill and Gulf*; Kenneth W. Nelson to All ASARCO Managers, "Memorandum: Environmental Lead Problem at El Paso Smelter," March 27, 1972, *Yoss et al. v. Bunker Hill and Gulf*; and James L. McNeil, MD, "Deposition," October 20, 1981, *Yoss et al. v. Bunker Hill and Gulf*.
36. Telephone interview with Philip Landrigan, January 6, 2006.
37. Christian Warren, "The Screaming Epidemic," in *Brush with Death*, 178–202.
38. Telephone interview with Philip Landrigan, January 6, 2006.
39. Ibid.
40. "Human Lead Absorption—Texas," *MMWR*, December 8, 1973, 405–407.
41. "Grim Days for El Paso," *Time*, March 27, 1972, http://www.time.com/time/magazine/article/0,9171,910286,00.html?promoid=googlep, accessed January 24, 2011.
42. Mike Cochran, "El Paso Fights Lead Poisoning," *Washington Post*, March 26, 1972, A1.
43. "El Paso Mayor Seeks Federal Aid in Area Hit by Lead Poisoning," *Wall Street Journal*, March 24, 1972, 7.
44. "Human Lead Absorption—Texas," *MMWR*, December 8, 1973, 405–407.
45. Nelson to All ASARCO Managers, "Memorandum," March 27, 1972, *Yoss et al. v. Bunker Hill and Gulf*; and ASARCO Inc., "News Release," May 15, 1972, KCA, Box 112 Health Department Director Issues Files, Folder 11 Lead Poisoning 1972–1979. In their news release on the consent decree entered into with the city of El Paso and the state of Texas, ASARCO asserted that "the current operation of the plant is not endangering the health of El Paso residents. Publicity and actions by city officials during the trial implied that the health of the entire City of El Paso was endangered. The medical examinations and therapy called for in Article IV of the decree were voluntarily initiated by Asarco in February when the elevated blood lead levels in Smeltertown children, apparently due to contaminated soil, were first discovered."
46. Jose Roman, MD, Jose D. Alva, MD, Jorge C. Magna, MD, and James L. McNeil, MD, "Minutes: Meeting of the Lead Surveillance Committee of the El Paso County Medical Society held on June 28, 1972" (n.d.), obtained from Lin Nelson, professor, Evergreen University.
47. Bertram W. Carnow, Bernard F. Rosenblum, and V. Carnow, "Unsuspected Community Lead Intoxication and Emissions from a Smelter: The El Paso Study," Paper presented at the Air Pollution Control Association 66th Annual Meeting, Chicago, June 26, 1973.
48. Sheehy to Files, "Memorandum," October 2, 1974, *Yoss et al. v. Bunker Hill and Gulf*. The notes summarize several internal industry meetings which were part of Gulf Resources and Bunker Hill's response to the childhood lead poisoning problem discovered near the Bunker Hill smelter. One of the meetings was between John J. Sheehy, an attorney at Rogers & Wells; Robert J. Muth, ASARCO's associate general counsel; and Kenneth Nelson, ASARCO's director of environmental services. The notes describe ASARCO's response to the El Paso crisis.

49. Ibid.; and Nelson to All ASARCO Managers, "Memorandum," March 27, 1972, *Yoss et al. v. Bunker Hill and Gulf.*
50. Sheehy to Files, "Memorandum," October 2, 1974, *Yoss et al. v. Bunker Hill and Gulf.*
51. Ibid.
52. P. J. Landrigan, S. H. Gehlbach, B. F. Rosenblum, J. M. Shoults, R. M. Candelaria, W. F. Barthel, J. A. Liddle, A. L. Smrek, N. W. Staehling, and J. F. Sanders. "Epidemic Lead Absorption Near an Ore Smelter: the Role of Particulate Lead," *New England Journal of Medicine* 292 (January 16, 1975): 123–129.
53. Sheehy to Files, "Memorandum," October 2, 1974, *Yoss et al. v. Bunker Hill and Gulf.*
54. Ibid.
55. "American Smelting Gets Double-Barreled Order Over El Paso Smelter," *Wall Street Journal*, May 15, 1972, 10; and Sheehy to Files, "Memorandum," October 2, 1974, *Yoss et al. v. Bunker Hill and Gulf.*
56. Sheehy to Files, "Memorandum," October 2, 1974, *Yoss et al. v. Bunker Hill and Gulf.*
57. "El Paso Smelter Still Poses Lead-Poisoning Peril to Children in Juarez," *New York Times*, November 28, 1977, 22.
58. "American Smelting Gets Double-Barreled Order"; and "El Paso Smelter Still Poses Lead-Poisoning Peril," 22.
59. Editorial, *El Paso Herald Post*, May 15, 1972; and Editorial, *El Paso Times*, May 15, 1972, King County Archives, hereafter KCA, Box 112 Health Department Director Issue Files, Folder 11 Lead Poisoning, 1972–79.
60. Sheehy to Files, "Memorandum," October 2, 1974, *Yoss et al. v. Bunker Hill and Gulf.*
61. Monica Perales, "Smeltertown: A Biography of a Mexican American Community, 1880–1973" (Ph.D. dissertation, Stanford University, December 2003, UMI No. 3111778), 279–323; and Monica Perales, *Smeltertown: Making and Remembering a Southwest Border Community* (Chapel Hill: University of North Carolina Press, 2010).
62. Perales, "Smeltertown"; and Perales, *Smeltertown*.
63. "High Levels of Lead in Texans Is Linked to El Paso Smelter," *New York Times*, December 18, 1973, 27.
64. Sheehy to Files, "Memorandum," October 2, 1974, *Yoss et al. v. Bunker Hill and Gulf.*
65. Bernard F. Rosenblum, MD, Director, El Paso City-County Health Department, to Philip J. Landrigan, MD, May 16, 1973, obtained from Lin Nelson, professor, Evergreen State University; and Sheehy to Files, "Memorandum," October 2, 1974, *Yoss et al. v. Bunker Hill and Gulf.*
66. ILZRO, "EH Agreement, LH-208," January 1, 1973. Contract for McNeil's study signed by James L. McNeil and Schrade F. Radke, Executive Vice President and Director of Research for ILZRO, *Yoss et al. v. Bunker Hill and Gulf*, Files of Paul Whelan, Plaintiff's attorney; D. R. Lynam and J. F. Cole, "Steering Committee Meeting LH-208, Epidemiologic Study of a Lead Contaminated Area," February 11, 1974, *Yoss et al. v. Bunker Hill and Gulf*, Files of Paul Whelan, Plaintiff's attorney; and ILZRO, "Comments on Testimony at the Hearing," November 21, 1975, *Yoss et al. v. Bunker Hill*, Files of Paul Whelan, Plaintiff's attorney.
67. James L. McNeil, "Evaluation of Long Term Effects of Elevated Blood Lead in Asymptomatic Children," Proposal to the International Lead Zinc Research Organization, August 1972, rev. November, 1972, *Yoss et al. v. Bunker Hill and Gulf*, Files of Paul Whelan, Plaintiff's attorney.
68. Ibid.
69. On Barltrop and Chisholm's funding from ILZRO, see, for example, Donald Barltrop to Jerome F. Cole, "LH-261, Management of Children with Abnormal Lead Exposure,"

January 23, 1978, *Yoss et al. v. Bunker Hill and Gulf*, Files of Paul Whelan; on Barltrop and Sachs consulting for ILZRO, see Jerome F. Cole to Dr. Jerome A. Lackner "Air Quality Advisory Committee, November 24, 1976, *Yoss et al. v. Bunker Hill and Gulf*, Files of Paul Whelan; on Chisholm's view that lead paint was the most important exposure to environmental lead, see Gerald Markowitz and David Rosner, *Lead Wars: the Politics of Science and the Fate of America's Children* (Berkeley: University of California Press, 2013), 98–99; and on Sachs's view that exposure to lead in air was not harming children's health, see ILZRO, "Comments on Testimony," November 21, 1975, *Yoss et al. v. Bunker Hill and Gulf*, Files of Paul Whelan, Plaintiff's attorney.

70. Lynam and Cole, "Steering Committee Meeting LH-208," February 11, 1974, *Yoss et al. v. Bunker Hill and Gulf.*
71. ILZRO, "EH Agreement, LH-208."
72. Bernard F. Rosenblum, MD, Director, El Paso City-County Health Department, to Philip J. Landrigan, MD, May 16, 1973, obtained from Lin Nelson, professor, Evergreen State University.
73. Telephone interview with Philip Landrigan, January 6, 2006.
74. Philip Landrigan, R. H. Whitworth, R. W. Baloh, N. W. Staehling, W. F. Barthel, and B. F. Rosenblum, "Neuropsychological Dysfunction in Children with Chronic Low-Level Lead Absorption," *Lancet* 305 (March 29, 1975): 708–712.
75. McNeil, "Deposition," October 20, 1981, *Yoss et al. v. Bunker Hill and Gulf.*
76. James L. McNeil and J. A. Ptasnik, "Evaluation of Long-term Effects of Elevated Blood Lead Concentrations in Asymptomatic Children," Paper presented at the Symposium on Recent Advances in the Assessment of the Health Effects of Environmental Pollution, World Health Organization, Paris, France, June 26, 1974, *Yoss et al. v. Bunker Hill and Gulf*; and Philip J. Landrigan, MD, to Hon. Troy Webb, Assistant Attorney General, State of Texas, April 23, 1973, obtained from Lin Nelson, professor, Evergreen State College.
77. Hill and Knowlton, Inc., "Draft Release: El Paso Study Finds No Subtle Damage to Children Exposed to High Lead Levels," June 12, 1974, *Yoss et al. v. Bunker Hill and Gulf*, Files of Paul Whelan, Plaintiff's attorney.
78. James D. Callaghan, Hill & Knowlton, Inc., to Dr. Jerome F. Cole, ILZRO, "Memorandum: Release on El Paso Study," June 12, 1974, *Yoss et al. v. Bunker Hill and Gulf*, Files of Paul Whelan, Plaintiff's attorney.
79. Callaghan to Robinson, "Memorandum: Further Distribution of the McNeil Study," July 1, 1974, *Yoss et al. v. Bunker Hill and Gulf.*
80. McNeil and Ptasnik, "Evaluation of Long-Term Effects."
81. Jerome F. Cole to Dr. James L. McNeil, "LH-208-Epidemiologic Study of a Lead Contaminated Area," July 23, 1974, *Yoss et al. v. Bunker Hill and Gulf*, Files of Paul Whelan, Plaintiff's attorney. Cole wrote to McNeil, "I feel a bit uneasy that a full statistical analysis of the results of the study have not been completed as yet. I think that you must make an effort to get this matter finalized as soon as possible or we will be open to criticism from the Steering Committee and outsiders because of the considerable amount of publicity the study has already received."
82. J. Julian Chisolm to Jerome F. Cole, July 26, 1974, *Yoss et al. v. Bunker Hill and Gulf*, Files of Paul Whelan, Plaintiff's attorney.
83. Philip E. Enterline to Jerome F. Cole, September 4, 1974, *Yoss et al. v. Bunker Hill and Gulf*, Files of Paul Whelan, Plaintiff's attorney.
84. James L. McNeil to Jerome F. Cole, April 2, 1975, *Yoss et al. v. Bunker Hill and Gulf*, Files of Paul Whelan, Plaintiff's attorney.

85. Cole to McNeil, "LH-208-Epidemiologic Study," July 23, 1974, *Yoss et al. v. Bunker Hill and Gulf*; and Jerome F. Cole to James L. McNeil, "Memorandum: Subject-LH-208, Epidemiologic Study of a Lead Contaminated Area," November 30, 1977, *Yoss et al. v. Bunker Hill and Gulf*, Files of Paul Whelan, Plaintiff's attorney.
86. McNeil, "Deposition," October 20, 1981, 11, *Yoss et al. v. Bunker Hill and Gulf*.
87. Office of Research and Development, "Air Quality Criteria for Lead 1977," appendix E, "Abstract of a Review of Three Studies," E1–E4.
88. P. J. Landrigan and E. L. Baker, "Letter: Child Health and Environmental Lead," *British Medical Journal*, March 26, 1977, 836; and James L. McNeil, MD to Dr. Joan R. Mortimer, May 24, 1974, *Yoss et al. v. Bunker Hill and Gulf*.
89. Cole to McNeil, "Memorandum: Subject-LH-208," November 30, 1977, *Yoss et al. v. Bunker Hill and Gulf*.
90. Callaghan to Robinson, "Memorandum: Further Distribution of the McNeil Study," July 1, 1974, *Yoss et al. v. Bunker Hill and Gulf*; ILZRO, "Comments on Testimony," November 21, 1975, *Yoss et al. v. Bunker Hill and Gulf*.
91. McNeil, "Deposition," October 20, 1981, *Yoss et al. v. Bunker Hill and Gulf*.
92. Office of Research and Development, "Air Quality Criteria for Lead 1977," appendix E, "Abstract of a Review of Three Studies," E4.
93. Sheehy to Files, "Memorandum," October 2, 1974, *Yoss et al. v. Bunker Hill and Gulf*.
94. "El Paso Smelter Still Poses Lead-Poisoning Peril," 22.
95. Drexler, *Study on the Source*. Drexler cites "data from the State of Texas" for these lead and arsenic figures; F. Diaz-Barriga, L. Batres, J. Calderon, A. Lugo, L. Galvao, I. Lara, P. Rizo, M. E. Arrovave, and R. McConnell, "The El Paso Smelter 20 Years Later: Residual Impact on Mexican Children," *Environmental Research* 74 (1997): 11–16; and R. Blumenthal, "Copper Plant Illegally Burned Hazardous Waste, EPA Says," *New York Times*, October 11, 2006, A18.
96. Sheehy to Files, "Memorandum," October 2, 1974, *Yoss et al. v. Bunker Hill and Gulf*.

CHAPTER 4 BUNKER HILL

1. Charles F. Barber to Frank G. Woodruff, March 29, 1972, *Edna Grace Yoss, Raymond Hans Yoss, Arlene Mae Yoss, Richard A. McCartney, Christina M. McCartney, Paula A. McCartney, Raymond E. Dennis, and Harley Dennis, by their guardian ad litem, Brian J. Linn, Plaintiffs, v. Bunker Hill Company, a Delaware corporation; Gulf Resources & Chemical Corporation, a Delaware corporation; Defendants*, No. CIV-77-2030, United States District Court for the District of Idaho, hereafter, *Yoss et al. v. Bunker Hill and Gulf*; and Kenneth W. Nelson to All ASARCO Managers, "Internal ASARCO Memorandum: Environmental Lead Problem at El Paso Smelter," March 27, 1972, Report titled "El Paso Smelter Environmental Lead Problem" attached, *Yoss et al. v. Bunker Hill and Gulf*.
2. Others have written the general outlines of the story of the lead poisoning of Silver Valley's children. Kathryn Aiken addresses the issue in *Idaho's Bunker Hill: The Rise and Fall of a Great Mining Company, 1885–1981* (Norman: University of Oklahoma Press, 2005); and John Osborn, a Spokane physician and environmentalist, has written extensively on his website about environmental issues in the Silver Valley, including the history of lead poisoning. See for example *Purging Mining's Poisons: Cleaning up the Coeur d'Alene Basin*, November 14, 2000, http://waterplanet.ws/documents/000314/. There have also been accounts of the lead poisoning epidemic and company response in numerous newspaper articles; see, for example, Robert McClure, "Town Split over Direction of Cleanup," *Seattle Post-Intelligencer*, October 21, 2001, http://www.seattlepi.

com/local/41086_silver02.shtml, accessed July 14, 2010. These are among the sources that have provided background and context for this chapter. My aim is to delve deeply into the public health story and to provide a detailed accounting of events and issues using primary sources. With respect to "worst community lead poisoning disaster," see William H. Foege to Edward S. Gallagher, "Letter," April 8, 1980, *Yoss et al. v. Bunker Hill and Gulf.*

3. According to James Halley, a Bunker Hill executive, children were not told they were part of a screening program. See James H. Halley, "Deposition, Volume II," January 21, 1981, 51, *Yoss et al. v. Bunker Hill and Gulf.* Dr. Panke, in deposition, said that he thought that parents knew their children were participating in a screening, but he did not know what parents of patients screened at the Doctor's Clinic and several grade schools in the valley were told. See Ronald K. Panke, "Deposition," April 23, 1981, 76–77, *Yoss et al. v. Bunker Hill and Gulf.* Elsewhere in the same deposition he states that parents of children who attended or formerly attended Silver King were told in a letter that their children were being asked to participate in a lead health screening. Ibid., 73. At the next deposition he was asked by lawyers in the Yoss case to produce the letter but was unable to do so. See Ronald K. Panke, "Deposition," August 27, 1981, 4, *Yoss et al. v. Bunker Hill and Gulf.* Panke also noted that he did not provide the results to the school where children's urine lead levels were highest. Additionally, he could not recall whether the parents of a child with a high urine lead level were notified. Ibid., 8–10. In a letter to Bunker Hill describing the results of the screening, Panke notes that there was not "a single instance of parental disapproval or concern that we were searching for a lead problem." See Ronald K. Panke to Eugene Baker, "Memorandum," July 10, 1972, Exhibit A, *Yoss et al. v. Bunker Hill and Gulf.*

4. See David Michaels, *Doubt Is Their Product: How Industry's Assault on Science Threatens Your Health* (New York: Oxford University Press, 2008).

5. See "Deposition Exhibit M, M-1, M-2, M-4," n.d., *Yoss et al. v. Bunker Hill and Gulf.* This is a compilation of raw data on urine lead concentrations from Panke's study of Silver Valley children from June of 1972. Final results are summarized in Panke to Baker, "Memorandum," July 10, 1972, Exhibit A.

6. Panke, "Deposition," April 23, 1981; James H. Halley "Deposition," January 21, 1980, *Yoss et al. v. Bunker Hill and Gulf*; and Halley, "Deposition, Volume II," January 21, 1981, 51–55.

7. Halley "Deposition," January 21, 1980, 51.

8. L. A. Blanksma, "Letter to the Editor: The Resolution on Childhood Lead Poisoning," *American Journal of Public Health* 60 (July 1970): 1191–1192; and J. J. Chisolm, "Increased Lead Absorption: Toxicological Considerations," *Pediatrics* 48 (September 1971): 349–352.

9. J. L. Steinfeld, "The Surgeon General's Policy Statement on Medical Aspects of Childhood Lead Poisoning" (U.S. Department of Health and Human Services, August 1971).

10. Panke, "Deposition," April 23, 1981, 28–42.

11. Gil Camm, clinic manager, Doctor's Clinic, "Deposition," May 13, 1980, 22, 38–39 *Yoss et al. v. Bunker Hill and Gulf.*

12. John J. Sheehy, attorney, Rogers & Wells to Files, "Memorandum: Gulf Resources & Chemical Corporation—ILZRO," October 2, 1974, *Yoss et al. v. Bunker Hill and Gulf.* In addition to the financial relationship between Bunker Hill and the Doctor's Clinic, the clinic and the company were also represented by the same Kellogg-based law firm. Dr. Panke also had a number of professional relationships related to Bunker Hill and the lead industry. For example, he was a consultant to Bunker's law firm, particularly

with respect to medico-legal matters involving Bunker Hill; see Camm, "Deposition," May 13, 1980, 4–5, 36. Panke was also Bunker Hill's medical consultant to the Shoshone Project; see Panke, "Deposition," April 24, 1981, 24, *Yoss et al. v. Bunker Hill and Gulf*. He also consulted for ILZRO and was sent by the trade association to a California Air Resources Board hearing when the state's air standard for lead was being reconsidered. At the hearing, he contradicted testimony given by Dr. Philip Landrigan on studies conducted on Silver Valley children and said that as a community physician, "I never saw any sick kids." See Panke, "Deposition," April 22, 1981, 133–134, *Yoss et al. v. Bunker Hill and Gulf*; and State of California, Air Resources Board, "In the Matter of: Reconsideration of the Ambient Air Quality Standard for Lead, Comments of Ronald K. Panke," November 3 and 4, 1975, *Yoss et al. v. Bunker Hill and Gulf*.
13. A methodologically sound approach would have included some type of random sampling or, at the very least, a clearly specified and scientifically justified approach to sampling.
14. Panke to Baker, "Memorandum," July 10, 1972.
15. "Deposition Exhibit M, M-1, M-2, M-4." This is a compilation of raw data on urine lead concentrations from Panke's study of Silver Valley children's urine lead levels from June of 1972.
16. Robert A. Kehoe, "Metabolism of Lead under Abnormal Conditions," *Archives of Environmental Health* 8 (1964): 235–243, at 236.
17. J. Julian Chisolm, "Deposition," August 28, 1981, 50, *Yoss et al. v. Bunker Hill and Gulf*.
18. "Deposition Exhibit M, M-1, M-2."
19. Panke to Baker, "Memorandum," July 10, 1972.
20. Panke, "Deposition," August 27, 1981, 8–10.
21. Steve Hurst to Larry M. Belmont, "Memorandum: Meeting with Silver King Elementary School, Smelterville, September 12, 1974," September 17, 1974, *Yoss et al. v. Bunker Hill Gulf*, Files of Paul Whelan, Plaintiff's attorney.
22. Steinfeld, "Surgeon General's Policy Statement."
23. Based on the available evidence, it does not appear that Bunker Hill/Gulf conducted any follow-up studies to Panke's study of urine lead levels.
24. Gene Baker, "Deposition," January 19, 1981, 47–50, *Yoss et al. v. Bunker Hill and Gulf*; and Michael O. Varner to Gene M. Baker, "Letter," July 16, 1973, Plaintiff's Exhibit 204, *Yoss et al. v. Bunker Hill and Gulf*, Files of Paul Whelan, Plaintiff's attorney.
25. Gene Baker to Art Lennon, "Memorandum," July 6, 1973, as read and discussed in James H. Halley, "Deposition Vol. II," January 22, 1981, 99–108, *Yoss et al. v. Bunker Hill and Gulf*.
26. Gene Baker, "Confidential: Events Pertaining to Lead Health Problem in Kellogg," undated report written in collaboration with attorneys in 1979 or 1980, *Yoss et al. v. Bunker Hill and Gulf*.
27. Keith R. Long, "Production and Disposal of Mill Tailings in the Coeur d'Alene Mining Region, Shoshone County, Idaho; Preliminary Estimates," U.S. Department of the Interior, US Geological Survey, Open File Report 98–595 (1998). This estimate is in 1997 dollars.
28. Nicholas A. Casner, "Leaded Waters: A History of Mining Pollution on the Coeur d'Alene River in Idaho, 1900–1950" (Master's thesis, Boise State University, 1989).
29. Long, "Production and Disposal of Mill Tailings."
30. "15% Pay Cuts at Bunker Hill," *New York Times*, September 4, 1981, D3; and Richard D. James, "Ripple Effect Leads to Economic Distress in Idaho's Silver Valley," *Wall Street Journal*, October 8, 1981, 1.

31. Aiken, *Idaho's Bunker Hill*.
32. Bob Woodward and Carl Bernstein, "Agency Continues to Drag Feet on Firm Tied to Bugging Case," *Washington Post*, November 4, 1972, A3.
33. Robert L. Jackson, "2 GOP Fundraisers Have Big Stake in Race," *Los Angeles Times*, October 3, 1972, A10.
34. Bob Woodward and Carl Bernstein, "$100,000 Gift to Nixon Campaign Is Traced to Texas Corporation," *Washington Post*, October 6, 1972, A1.
35. Woodward and Bernstein, "Agency Continues to Drag Feet."
36. Ibid.
37. Casner, "Leaded Waters."
38. Steven V. Roberts, "Idaho's 'Hard Rock' Miners: They Work Hard, Play Hard and Don't Save a Dime . . . ," *New York Times*, May 7, 1972, 34.
39. Gene M. Baker, "The Bunker Hill Company Environmental Control Update," December 7, 1974. Paper presented at the Northwest Mining Association Convention at Spokane, Washington, *Yoss et al. v. Bunker Hill and Gulf*.
40. Gene M. Baker, "Confidential: Pollution Control Talk," May 1, 1970, *Yoss et al. v. Bunker Hill and Gulf*.
41. Douglas I. Hammer, MD, Epidemiology Associate, Ecological Research Branch, Division of Health Effects Research, Department of Health Education and Welfare, to Terrell O. Carver, MD, Administrator of Health, Idaho Dept. of Health, "Letter," April 27, 1970, *Yoss et al. v. Bunker Hill and Gulf*. This letter includes appendices that provide data on heavy metal concentrations in hair in East Helena, Helena, Bozeman, Anaconda, and Kellogg.
42. George Dekan to Jim Kimball, "Memorandum: Kellogg History," February 11, 1974, *Yoss et al. v. Bunker Hill and Gulf*.
43. Environmental Protection Agency, "Health Hazards of Lead" (1972), *Yoss et al. v. Bunker Hill and Gulf*, Files of Paul Whelan, Plaintiff's attorney.
44. Committee on Biologic Effects of Atmospheric Pollutants, *Lead: Airborne Lead in Perspective* (Washington, DC: National Academy of Sciences, 1972), 211.
45. H. Dale Henderson, Inc., "Fact File, The Bunker Hill Company, First Draft," January 1975, Yoss et al. v. Bunker Hill and Gulf. Surface temperature inversions occur when there is cool air at ground level and warmer air aloft. Warm air above traps air pollutants near the surface, impacting air quality. See "What Are Temperature Inversions?" U.S. Department of Commerce, National Oceanic and Atmospheric Administration, at http://www.wrh.noaa.gov/slc/climate/TemperatureInversions.php.
46. Environmental Services Staff to James A. Bax, Director State of Idaho, Department of Health and Welfare, "Memorandum: Status Report on Lead Emissions from Bunker Hill Company," October 9, 1974, EPA Superfund Records, Bunker Hill Site File (hereafter, BHSF), 1.3.1.v4.
47. Earl H. Bennett, "A History of the Bunker Hill Superfund Site, Kellogg, Idaho," Prepared for the Northwest Metals Conference, Spokane, Washington, April 9, 1994; and Aiken, *Idaho's Bunker Hill*.
48. Aiken, *Idaho's Bunker Hill*, 174.
49. Baker, "Confidential: Pollution Control Talk."
50. James A. Bax, Director State of Idaho Department of Health and Welfare, to Mr. Frank G. Woodruff, "Letter," July 17, 1974, *Yoss et al. v. Bunker Hill and Gulf*.
51. Gene M. Baker to R. H. Allen, "Memorandum: Bunker Hill EPA Study," March 13, 1980, *Yoss et al. v. Bunker Hill and Gulf*; Aiken, *Idaho's Bunker Hill*, 189; and "Bunker Hill and EPA Reach Agreement, Settlement Ends Four Years of Litigation," *Kellogg Evening News*, June 8, 1979, 1.

52. Unknown Author, NIOSH presentation, "Bunker Hill: Seattle Meeting," August 13, 1975, Yoss et al. v. Bunker Hill and Gulf, Files of Paul Whelan, Plaintiff's attorney; Bunker Hill, "Data Sheet, Authority for Expenditure, Project Data, 1974 Budget Year," January 28, 1974, Yoss et al. v. Bunker Hill and Gulf, Files of Paul Whelan, Plaintiff's attorney; and Baker, "Deposition," January 19, 1981, 99–102.
53. On prices, see "Annual Average Metal Prices—1910 to 1974," *Engineering and Mining Journal*, March 1975, 89, Yoss et al. v. Bunker Hill and Gulf. Silver prices had increased from 177.08 cents/oz in 1970 to 470.79 cents/oz in 1974; lead from 15.61 cents/lb in 1970 to 22.53 cents/lb in 1974; and zinc from 15.31 cents/lb to 35.94 cents/lb between 1970 and 1974. On pressure from Gulf, see Aiken, *Idaho's Bunker Hill*, 180–181. Notes from a conversation with Philip Landrigan have Ian von Lindern stating that after the fire damaged the baghouse and the company found they could not operate efficiently with half a baghouse, the "proper move would have been to shut down temporarily for repairs." See Philip Landrigan to the Record, "Lead Emissions Patterns—Kellogg, Idaho, Smelter," July 18, 1975, Yoss et al. v. Bunker Hill and Gulf, files of Paul Whelan, Plaintiff's attorney. On the production levels, see George M. Dekan, "Memorandum: Bunker Hill Lead History," (n.d., received by Environmental Services February 6, 1974), Yoss et al. v. Bunker Hill and Gulf. On the baghouse operation, see Baker, "Confidential: Events Pertaining to Lead Health Problem in Kellogg."
54. Baker, "Confidential: Events Pertaining to Lead Health Problem in Kellogg"; Bunker Hill Company, "Lead Smelter Emission Record, 1971–1977," n.d., Yoss et al. v. Bunker Hill and Gulf, Exhibit #32; and Landrigan to the Record, "Lead Emission Patterns."
55. Baker, "Confidential: Events Pertaining to Lead Health Problem in Kellogg."
56. "Five-Year Summary" (Bunker Hill Company financial data), n.d., Plaintiff's Exhibit, No. 227, Yoss et al. v. Bunker Hill and Gulf, Files of Paul Whelan, Plaintiff's attorney.
57. Dekan to Kimball, "Memorandum: Kellogg History"; and Ian von Lindern to Bob Olson, "Memorandum: Emissions of Lead Particulate from the Bunker Hill Company Smelter Complex and the Associated Contamination of the Local Environment," May 17, 1974, Yoss et al. v. Bunker Hill and Gulf.
58. Dekan, "Memorandum: Bunker Hill Lead History."
59. Dekan to Kimball, "Memorandum: Kellogg History."
60. Jerry Cobb and Annabelle Rose to Larry Belmont, Hillie Barry, Jim Kimball, and Dr. Reeds, "Memorandum: Taylor Family [pseudonym]—Lead Poisoning Case," April 25, 1974, Yoss et al. v. Bunker Hill and Gulf; and Centers for Disease Control, "Protocol: Epidemiologic Investigation of Human Lead Absorption near a Smelter-Kellogg, Idaho" (n.d., circa June 1974), Yoss et al. v. Bunker Hill and Gulf.
61. Thomas O. Reeds, "Deposition," May 12, 1980, 17–18, Yoss et al. v. Bunker Hill and Gulf; and CDC, "Protocol."
62. M. G. Weisskopf, N. Jain, H. Nie, D. Sparrow, P. Vokonas, J. Schwartz, and H. Hu, "A Prospective Study of Bone Lead Concentration and Death from All Causes, Cardiovascular Diseases, and Cancer in the Department of Veterans Affairs Normative Aging Study," *Circulation* 120 (2009): 1056–1064.
63. H. Hu and M. Hernandez-Avila, "Invited Commentary—Lead, Bones, Women, and Pregnancy—The Poison Within?" *American Journal of Epidemiology* 156 (2002): 1088–1091; and J. R. Pilsner, H. Hu, A. Ettinger, B. N. Sanchez, R. O. Wright, D. Cantonwine, A. Lazarus, et al., "Influence of Prenatal Lead Exposure on Genomic Methylation of Cord Blood DNA," *Environmental Health Perspectives* 117 (2009), 1466–1471.
64. Cobb and Rose to Belmont, Barry, Kimball, and Reeds, "Memorandum."
65. CDC, "Protocol."

66. Reeds, "Deposition," May 12, 1980.
67. Cobb and Rose to Belmont, Barry, Kimball, and Reeds, "Memorandum."
68. James A. Bax, administrator, Department of Environmental and Community Services, to Gene Baker, Director Environmental Control, Bunker Hill, "Letter," May 22, 1974, *Yoss et al. v. Bunker Hill and Gulf.*
69. Gene M. Baker to Land Owners in the Kellogg Area, "Letter," April 29, 1974, *Yoss et al. v. Bunker Hill and Gulf.*
70. Bob Cutchins [pseudonym], "Deposition," July 27, 1981, 28–36, *Yoss et al. v. Bunker Hill and Gulf.*
71. M. W. Aiken to R. W. Crosser, February 3, 1975, *Yoss et al. v. Bunker Hill and Gulf*; and Cutchins, "Deposition," July 27, 1981, 38.
72. Cutchins, "Deposition," July 27, 1981, 38.
73. Ibid; and Dr. Brock to Jim Dodds, "Memorandum: Results of Lead Tests on Cutchins Family [pseudonym]," September 20, 1974, *Yoss et al. v. Bunker Hill and Gulf.*
74. Shelly Long, legal assistant, "Deposition Summary of Philip J. Landrigan, M.D.," January 17, 1980, *Yoss et al. v. Bunker Hill and Gulf*, Files of Paul Whelan, Plaintiff's attorney.
75. CDC, "Protocol."
76. Robert E. Brown to Robert H. Allen, "Telegram," September 4, 1970, *Yoss et al. v. Bunker Hill and Gulf.*
77. P. J. Landrigan, E. L. Baker, R. G. Feldman D. H. Cox, K. V. Eden, W. A. Orenstein, J. A. Mather, A. J. Yankel, and I. H. Lindern, "Increased Lead Absorption with Anemia and Subclinical Neuropathy in Children Near a Lead Smelter," n.d., *Yoss et al. v. Bunker Hill and Gulf*, draft of paper published as P. J. Landrigan, E. L. Baker, R. G. Feldman, D. H. Cox, K. V. Eden, W. A. Orenstein, J. A. Mather, A. J. Yankel, and I. H. Lindern, "Increased Lead Absorption with Anemia and Slowed Nerve Conduction in Children Near a Lead Smelter," *Journal of Pediatrics* 89 (1976): 904–910.
78. Cutchins, "Deposition," July 27, 1981, 107.
79. Wafford Conrad and Cheney Cowles, "Lead Poison Study Hit," *Spokane Daily Chronicle*, September 6, 1974, 1; and Barbara Wilkins, "Lead Poisoning Threatens the Children of an Idaho Town," *People*, September 30, 1974, http://www.people.com/people/archive/article/0,,20064515,00.html, accessed January 24, 2011.
80. Camm, "Deposition," May 13, 1980, 22.
81. "Lead Poisoning Takes Heavy Toll in Idaho Mining Town," *Sarasota Herald Tribune*, September 26, 1974, A9.
82. Conrad and Cowles, "Lead Poison Study Hit."
83. Wilkins, "Lead Poisoning Threatens the Children."
84. "Tri-Citians Suing Bunker Hill Cited as Unfit Parents," *Tri-City Herald*, October 9, 1981, 8; and S. M. Moodie, E. K. Tsui, and E. K. Silbergeld, "Community and Family-Level Factors Influence Care-Giver Choice to Screen Blood Lead Levels of Children in a Mining Community," *Environmental Research* 110 (2010): 484–496.
85. Centers for Disease Control and Prevention, "Managing Elevated Blood Lead Levels among Young Children: Recommendations from the Advisory Committee on Childhood Lead Poisoning Prevention," March, 2002, 84; and David C. Bellinger, "Lead," *Pediatrics* 113, Suppl. 4 (2004): 1016–1022.
86. "Lead Poisons Most of Town's Children," *Los Angeles Times*, September 6, 1974, A1.
87. Unknown Author, "Confidential handwritten document" (n.d.), BHSF 1.2.3. The document, which contains notes and calculations on the cost of lead poisoning in the Silver Valley, has been attributed to Bunker Hill executives or their lawyers and was used as evidence in *Yoss et al. v. Bunker Hill and Gulf.*

NOTES TO PAGES 89-91

88. Jerome F. Cole to Mr. G. Baker, "Letter," September 10, 1974, *Yoss et al. v. Bunker Hill and Gulf*, Files of Paul Whelan, Plaintiff's attorney; Baker, "Confidential: Events Pertaining to Lead Health Problem in Kellogg"; Baker, "Deposition," January 19, 1981, 108; and Sheehy to Files, "Memorandum."
89. James D. Callaghan, Hill & Knowlton, Inc., to Philip E. Robinson, Lead Industries Association, Inc., "Memorandum: Idaho Lead Smelter Story," September 6, 1974, *Yoss et al. v. Bunker Hill and Gulf*, Files of Paul Whelan, Plaintiff's attorney.
90. Cole to Baker, "Letter," September 10, 1974.
91. Schrade F. Radke to Frank G. Woodruff, "Letter," October 10, 1974, Yoss et al. v. Bunker Hill and Gulf, Files of Paul Whelan, Plaintiff's attorney.
92. Baker, "Confidential: Events Pertaining to Lead Health Problem in Kellogg"; and Baker, "Deposition," January 19, 1981, 108.
93. Robert H. Allen, "Deposition," March 24, 1981, 78 and 87, *Yoss et al. v. Bunker Hill and Gulf*.
94. See, for example, Foege to Gallagher, "Letter," April 8, 1980.
95. Bunker Hill funded the Shoshone Project and paid the salary of its director, Glen Wegner; see "Memorandum of Agreement between J. H. Halley, President, The Bunker Hill Company, Dr. James A. Bax, Director, Department of Health and Welfare, State of Idaho and Glen Wegner, MD, JD, Private Attorney and Physician," Week of October 14, 1974, *Yoss et al. v. Bunker Hill and Gulf*. There were multiple public statements about the key findings of the Shoshone Project, that children had not been permanently harmed. See, for example, "Wegner 'Cautiously Optimistic' No Serious Harm Done by Lead," *Lewiston Morning Tribune*, July 28, 1975, A7; and Bill Richards, "Worried Parents Question Smelter's Effect on Children," *Washington Post*, October 7, 1979. A Gulf motion in the Yoss case referred to the "so-called lead problem in Kellogg"; see "Gulf Asks Curb on Lawsuit Publicity" (unknown newspaper, n.d.), Files of Paul Whelan, Plaintiff's lawyer. See statements to this effect from the State Health Department in "Kellogg Kids to Get Lead-in-Blood Exams" *Spokesman Review*, March 27, 1980, 1.
96. See Sheehy to Files, "Memorandum." Notes from a meeting of industry officials and lawyers state: "Bunker's desire to have ILZRO handle the referral and payment for treatment was, in part, based upon the difference in the Kellogg and El Paso problems. In El Paso, ASARCO was the only smelter in the area. Accordingly, they would have to bear responsibility for whatever damage occurred. In Kellogg, however, while Bunker is the only smelter, there are other mills in the area. It is not certain where the lead comes from; i.e., dust, soil, direct ingestion, etc."
97. Ibid.
98. Kent Swigard, "Lead Poisoning: Health Director Raps New Study," *Spokane Spokesman-Review*, September 27, 1974, 6.
99. Larry M. Belmont, director, Panhandle Health District I, "Memorandum," October 8, 1974, Plaintiff's Exhibit 5, *Yoss et al. v. Bunker Hill and Gulf*.
100. Ronald K. Panke to Jerome F. Cole, "Letter," October 30, 1974, *Yoss et al. v. Bunker Hill and Gulf*, Files of Paul Whelan, Plaintiff's attorney.
101. Belmont, "Memorandum"; and "Memorandum of Agreement between J. H. Halley, Dr. James A. Bax, and Glen Wegner," *Yoss et al. v. Bunker Hill and Gulf*.
102. See, for example, Gene M. Baker to Jerome F. Cole, "Letter," June 18, 1975, *Yoss et al. v. Bunker Hill and Gulf*, Files of Paul Whelan, Plaintiff's attorney. Baker asks for Cole's assistance in analyzing a study of Kellogg children's intelligence tests that was part of the Shoshone Project and promises to forward to ILZRO the names of the Kellogg children who had taken part in the tests.

103. Panke to Cole, "Letter."
104. Baker, "Confidential: Events Pertaining to Lead Health Problem in Kellogg."
105. G. R. to Frank [Woodruff], "Confidential Handwritten Notes: Information from Dr. Raymond Suskind, Director of Kettering Lab, University of Cincinnati," n.d., Yoss et al. v. Bunker Hill and Gulf.
106. "Suggested Names for Technical Steering Committee (suggested by Bunker Hill)," October 16, 1974, Yoss et al. v. Bunker Hill and Gulf.
107. Ted A. Loomis, "Testimony," attachment to C. J. Newlands to R. J. Muth, February 16, 1973, Environmental Protection Agency Site File, Commencement Bay/Nearshore Tideflats ASARCO Smelter Facility Site File (hereafter, ASDSF) 1.6.3.
108. "Memorandum of Agreement between J. H. Halley, Dr. James A. Bax, and Glen Wegner"; "Glen E. Wegner, Biographical Note," Nixon Presidential Library & Museum, http://www.nixonlibrary.gov/forresearchers/find/textual/central/smof/wegner.php, accessed January 21, 2011; and Baker, "Deposition," January 19, 1981, 112–114.
109. On the cost of the study and Wegner's salary, see "Memorandum of Agreement between J. H. Halley, Dr. James A. Bax, and Glen Wegner"; and for an example of Bunker Hill's request for internal study information, see G. M. Baker to Dr. Glen Wegner, "Memorandum: Request for Information Regarding Lead Problem," November 12, 1974, Yoss et al. v. Bunker Hill and Gulf, Files of Paul Whelan, Plaintiff's attorney.
110. Sandy Travis [pseudonym], Public Health Educator, Idaho Division of Health, "Notes," June 11, 1975, Yoss et al. v. Bunker Hill and Gulf, Files of Paul Whelan, Plaintiff's attorney.
111. Dale Henderson, Inc., Public Relations & Marketing, "Confidential Report," n.d., Yoss et al. v. Bunker Hill and Gulf.
112. "Memorandum of Agreement between J. H. Halley, Dr. James A. Bax, and Glen Wegner."
113. Ibid.
114. Steinfeld, "Surgeon General's Policy Statement."
115. Ibid.
116. Edward S. Gallagher, MD, State Health Officer, to Milton G. Klein, Director Idaho Department of Health and Welfare, "Memorandum: Proposals for Three Separate Areas of Study and a (Fourth) Course of Action of the Shoshone Lead Problem as It Relates to Human Health," February 1, 1980, Yoss et al. v. Bunker Hill and Gulf.
117. Dennis F. Brendel, vice president of environmental safety, Bunker Hill Company, to Dr. Edward Gallagher, "Memorandum: Grant to Fund Development of Health and Welfare Continuing Program to Monitor and Eliminate Potential Lead/Health Problems in the Silver Valley," March 24, 1980, Yoss et al. v. Bunker Hill and Gulf.
118. Unknown Author, "Matched Prenatal-Newborn Pairs—Blood Lead Levels, Shoshone County 1976–1978," November 1979, Deposition Exhibit H-7, Yoss et al. v. Bunker Hill and Gulf.
119. Dennis F. Brendel to Dr. Edward S. Gallagher, "Letter," May 5, 1980, Yoss et al. v. Bunker Hill and Gulf; and Coldevin Carlson, "Deposition," n.d., Yoss et al. v. Bunker Hill and Gulf.
120. Dr. Keith Dahlberg to Dr. Carlson, "Letter," January 14, 1975, Yoss et al. v. Bunker Hill and Gulf.
121. On claims to the contrary, for example, in response to a question from Bill Richards of the Washington Post, Bunker Hill stated that "Bunker Hill was periodically informed of progress [of the Shoshone Project], but was not involved in decision making." See "Questions Asked by Bill Richards (Washington Post) on September 20, 1979; Answers Supplied by Bunker Hill on September 21, 1979," Yoss et al. v Bunker Hill and Gulf, Files of Paul Whelan, Plaintiffs attorney. On protesting a disabled child's inclusion in the

study, see G. M. Baker to Dr. Glen Wegner, "Memorandum: Shoshone County Study," October 31, 1974, *Yoss et al. v. Bunker Hill and Gulf.*

122. Dr. Glenn Wegner, director, Shoshone Project, "Lead Report: An Open Letter to the People of Shoshone County," January 1975, *Yoss et al. v. Bunker Hill and Gulf.*
123. Ibid.
124. See Baker, "Bunker Hill Company Environmental Control Update"; and "Lead Reduced," *North Idaho Press,* July 18, 1975, 1. Also see Glen Wegner's statement about the company's efforts to reduce lead in the environment in Wegner, "Lead Report."
125. Ian von Lindern, environmental engineer, State of Idaho, Department of Health and Welfare, "File Note: Bunker Hill Inspection," July 29, 1975, BHSF 1.3.1.
126. Ian von Lindern to Lee Stokes, "Memorandum: Agency Credibility and Effectiveness in Dealing with the Bunker Hill Company," June 9, 1975, Yoss et al. v. Bunker Hill and Gulf.
127. Landrigan to the Record, "Memorandum: Lead Emission Patterns."
128. Scott W. Reed to Mr. John Lundin, "Memorandum: Taylor [pseudonym] v. Gulf Resources et al.," August 4, 1975, *Yoss et al. v. Bunker Hill and Gulf,* Files of Paul Whelan, Plaintiff's attorney. The letter describes a meeting with Ian von Lindern at which the Shoshone Project was discussed. Von Lindern shared his concerns and those of "four or five persons working for the Health Department."
129. Aiken, *Idaho's Bunker Hill,* 186–204.
130. H. Dale Henderson to Mr. James H. Halley, "Confidential Letter," January 8, 1975, *Yoss et al. v. Bunker Hill and Gulf.*
131. Gelb Marketing Research, "Confidential Report: The Bunker Hill Company as Seen by Citizens of Idaho and Spokane," Report prepared for the Bunker Hill Company and Ogilvy and Mather Inc., Dale Henderson, Inc., March, 1975, *Yoss et al. v. Bunker Hill and Gulf.*
132. H. Dale Henderson, Inc., "Confidential Report." The public relations firm wrote in the same report that the *Kellogg Evening News* "has been called a company paper. It isn't. It could become that with minor effort, however."
133. "Blood-Lead Levels Drop Near Smelter," *Los Angeles Times,* April 13, 1975, 2.
134. "Lead Reduced," *North Idaho Press,* July 18, 1975, 1.
135. "Wegner 'Cautiously Optimistic'."
136. Editorial, "Who's Watching Kellogg?" *Idaho State Journal,* July 30, 1975, A4.
137. Ibid.
138. Silver King Parent Teachers Association, "Did You Know?" Poster asking parents to vote to keep the school open (n.d., circa 1974), *Yoss et al. v. Bunker Hill and Gulf,* Files of Paul Whelan, Plaintiff's attorney.
139. See, for example, "Questions Asked by Bill Richards"; Brendel to Gallagher, "Memorandum"; and Richards, "Worried Parents Question Smelter's Effect."
140. Baker, "Bunker Hill Company Environmental Control Update."
141. Gallagher to Klein, "Memorandum: Proposals for Three Separate Areas of Study."
142. Landrigan, Baker, Feldman et al., "Increased Lead Absorption."
143. P. S. Gartside and R. K. Panke, "A Discussion Concerning the Significance of Results for Children Tested in the Shoshone Project," final draft for monograph, n.d., *Yoss et al. v. Bunker Hill and Gulf.*
144. "Statement by Shoshone Lead Project Technical Committee," September 4, 1975, *Yoss et al. v. Bunker Hill and Gulf.*
145. Panke, "Deposition," August 27, 1981, 19.
146. John T. Ashley, MD, state health officer, to Philip J. Landrigan, "Letter," (December 22, 1975), *Yoss et al. v. Bunker Hill and Gulf.* The letter in part concerns the state health

officer's objection to Landrigan including information in a paper he was submitting for publication from a personal communication with Dr. Reitan stating that children he tested had "subtle abnormalities in central function." Landrigan was told by the state health officer that they would not clear that sentence for publication. The health officer stated "we would be willing to discuss it further but first would like to examine the written communication from Dr. Reitan to you and probably have a meeting in Seattle with you and Dr. Reitan in order that all parties would understand the implications of the statement."

147. Panke, "Deposition," August 27, 1981, 20–22.
148. H. Dale Henderson to Mr. Gene M. Baker, "Letter," August 21, 1975, *Yoss et al. v. Bunker Hill and Gulf*, Files of Paul Whelan, Plaintiff's attorney.
149. Susan Lee, paralegal to Paul Whelan, Plaintiff's attorney, "Memorandum: Yoss v. Bunker Hill, conference with Sandy Travis [pseudonym], May 8, 1978," May 8, 1978, *Yoss et al. v. Bunker Hill and Gulf*, Files of Paul Whelan, Plaintiff's attorney.
150. "Attorney Claims Bunker Pressured State Senator" *Spokesman-Review*, October 1, 1981, 7.
151. See, for example, Richards, "Worried Parents."
152. Robert Bazell, NBC Nightly News, "Transcript: Lead and Learning," November 26, 1979, *Yoss et al. v. Bunker Hill and Gulf*.
153. Bill Curry, "Children Feared Harmed by Idaho Lead Smelting Plant," *Los Angeles Times*, September 21, 1980, 1.
154. "Statement of Philip Landrigan," Oversight—Clean Air Act Amendments of 1977, 80-H501–65; and Baker, "Confidential: Events Pertaining to Lead Health Problem in Kellogg."
155. State of Idaho, "Shoshone County Lead Study Ages 1–10, Table I," October 1980, *Yoss et al. v. Bunker Hill and Gulf*; and Gallagher to Klein, "Memorandum: Proposals for Three Separate Areas."
156. Committee on Environmental Health, "Policy Statement: Lead Exposure in Children: Prevention, Detection, and Management" 116 (October 2005): 1036–1046; and K. R. Mahaffey, J. L. Annest, and J. Roberts, "National Estimates of Blood Lead Levels: United States, 1976–1980—Association with Selected Demographic and Socioeconomic Factors," *New England Journal of Medicine* 307 (1982): 573–579.
157. State of Idaho, "Shoshone County Lead Study."
158. Brendel to Gallagher, "Letter: Grant to Fund Department of Health and Welfare."
159. "Statement of Philip Landrigan."
160. Edward S. Gallagher to Peter D. Drotman, MD, Bureau of Epidemiology, CDC, "Letter," March 14, 1980, *Yoss et al. v. Bunker Hill and Gulf*.
161. Fritz R. Dixon, Chief Bureau of Preventive Medicine, IDHW, to Dennis Brendel "Letter," March 19, 1980, *Yoss et al. v. Bunker Hill and Gulf*.
162. Ibid.; and Brendel to Gallagher, "Letter: Grant to Fund Department of Health and Welfare."
163. Edward S. Gallagher to Peter Drotman, "Letter," March 26, 1980, *Yoss et al. v. Bunker Hill and Gulf*.
164. Foege to Gallagher, "Letter," April 8, 1980.
165. Bob Mims, "Smokestacks Reminder of Lead Tragedy," *Sunday Oregonian*, April 12, 1981, C3.
166. Idaho Department of Health & Welfare, Bureau of Preventive Medicine, "Status of Blood Lead Determinations in Shoshone County," April 1980, *Yoss et al. v. Bunker Hill and Gulf*.

167. Centers for Disease Control, "Preventing Lead Poisoning in Young Children" October 1, 1991, http://wonder.cdc.gov/wonder/prevguid/p0000029/p0000029.asp#head007001001000000.
168. Idaho Department of Health and Welfare, "Status of Blood Lead Determinations."
169. "Kellogg Kids to Get Lead-in-Blood Exams."
170. Bazell, "Lead and Learning."
171. Shelly Long, "Deposition Summary, John A. Mather," deposition taken July 8, 1980, *Yoss et al. v. Bunker Hill and Gulf*, Files of Paul Whelan, Plaintiff's attorney. This is a summary of Dr. Mather's deposition rather than a verbatim transcript. Therefore the quote represents a summarized statement.
172. Fritz R. Dixon to Mr. Mel Ott or Dr. Dennis Brendel, "Memorandum," June 2, 1980, *Yoss et al. v. Bunker Hill and Gulf*; and Fritz R. Dixon to Dennis Brendel, "Letter," July 23, 1980, Deposition Exhibit 0–13, *Yoss et al. v. Bunker Hill and Gulf*.
173. Fritz R. Dixon to Ronald K. Panke, "Letter," June 20, 1980, Deposition Exhibit 0–15, *Yoss et al. v. Bunker Hill and Gulf*.
174. Brenda D. Townes, associate professor psychiatry and behavioral sciences, to Jerry Sells, Child Development and Mental Retardation Center, University of Washington, "Memorandum: Lead Study Patients," March 24, 1981, *Yoss et al. v. Bunker Hill and Gulf*, Files of Paul Whelan, Plaintiff's attorney. In addition to performing poorly on the Category Test, Townes reported that "6 of the children show evidence of mild to moderate impairment in adaptive abilities suggestive of organic brain dysfunction." She summarized: "as a group, these 11 children show subtle impairments in abilities beyond that which might be expected in a normal population. The potential effects of exposure to high lead concentration during childhood may be interacting with effects from trauma and disease to produce the subtle impairments seen in these 11 patients. It is not possible to determine with confidence, however, the relative contribution of disease, trauma, exposure to lead, and normal deviations in development in producing the mild to moderate impairments seen in this small group of children."
175. Jerome F. Cole to James L. McNeil, "Letter," October 8, 1975, *Yoss et al. v. Bunker Hill and Gulf*, Files of Paul Whelan, Plaintiff's attorney.
176. Donald R. Lyman, "Visit Report: Dr. Clark Cooper and Dr. William Gaffey," October 2, 1975, *Yoss et al. v. Bunker Hill and Gulf*, Files of Paul Whelan, Plaintiff's attorney.
177. ILZRO, "Comments on Testimony at the Hearing on Reconsideration of the California Ambient Air Quality Standard for Lead, California Air Resources Board, Sacramento Nov. 3 & 4, 1975," November 21, 1975, *Yoss et al. v. Bunker Hill and Gulf*, Files of Paul Whelan, Plaintiff's attorney; James L. McNeil, MD, "Deposition," October 20, 1981, 12–13, *Yoss et al. v. Bunker Hill and Gulf*; and Panke, "Deposition," August 27, 1981, 39.
178. ILZRO, "Comments on Testimony," 27–28.
179. Ibid., 4.
180. EPA, "Health Hazards of Lead."
181. Kenneth Bridbord and David Hanson, "A Personal Perspective on the Initial Federal Health-Based Regulation to Remove Lead from Gasoline," *Environmental Health Perspectives* 117 (August 2009): 1195–2101.
182. Gene M. Baker to Frank G. Woodruff, "Memorandum: Lead Emissions," March 18, 1976, *Yoss et al. v. Bunker Hill and Gulf*.
183. Ibid.
184. *Natural Resources Defense Council, Inc. v. Russell Train* 545 F.2d 320 (1976).
185. "EPA Lead Emission Rules Can't Be Met," *Coeur d'Alene Press*, June 8, 1979.

186. G. M. Baker to R. H. Allen, "Memorandum: EPA Lead Standard—Bunker Hill," May 15, 1979, *Yoss et al. v. Bunker Hill and Gulf.*
187. Environmental Protection Agency, "Bunker Limited Partnership Federal Implementation Plan for Lead," n.d., BHSF 1.3.7 v.1.
188. Baker to Allen, "Memorandum: Bunker Hill EPA Study," March 13, 1980.
189. *Lead Industries Association, Inc., Petitioner, v. Environmental Protection Agency, Respondent, Bunker Hill Company, Intervenor. St. Joe Minerals Corporation, Petitioner, v. Environmental Protection Agency, Respondent, Bunker Hill Company, Intervenor*—647 F.2d 1184 (1980).
190. Anthony J. Yankel, Ian H. von Lindern, and Stephen D. Walter, "The Silver Valley Lead Study: The Relationship between Childhood Blood Lead Levels and Environmental Exposure," *Journal of the Air Pollution Control Association* 27 (August 1977): 763–767.
191. Anthony J. Yankel, "Affidavit," April 29, 1980, *Yoss et al. v. Bunker Hill and Gulf*, Files of Paul Whelan, Plaintiff's attorney.
192. Ian H. von Lindern, "Letter: Re: Ambient Air Standard Docket 77–1," May 24, 1980, *Yoss et al. v. Bunker Hill and Gulf*, Files of Paul Whelan, Plaintiff's attorney.
193. *Lead Industries Association, Inc., Petitioner, v. Environmental Protection Agency, Respondent, Bunker Hill Company, Intervenor. St. Joe Minerals Corporation, Petitioner, v. Environmental Protection Agency, Respondent, Bunker Hill Company, Intervenor*—647 F.2d 1184 (1980).
194. Gene Baker, "Confidential: Background Notes for Meeting with Steelworkers," July 12, 1979, *Yoss et al. v. Bunker Hill and Gulf.*
195. Unknown author, "Notes to File, Re: Taylor [pseudonym] v. Bunker Hill," October 7, 1974, *Yoss et al. v. Bunker Hill and Gulf*, Files of Paul Whelan, Plaintiff's attorney.
196. NIOSH, "Interim Report, Environmental and Morbidity Phase: Bunker Hill Study," n.d., OSHA Docket #H-004.
197. United Steelworkers of America, "Post-Hearing Brief: Standard for Inorganic Lead," June 20, 1977, OSHA Docket #H-004.
198. Laurie Mercier, *Anaconda: Labor, Community, and Culture in Montana's Smelter City* (Urbana: University of Illinois Press, 2001), 201.
199. Bill Graves, "Bunker Hill Union May Take up Its Own Battle with OSHA," *Coeur d'Alene Press*, June 6, 1979, 1.
200. E. V. Howard to F. G. Woodruff, "Confidential Memorandum: P.R. Problem," October 1, 1979, *Yoss et al. v. Bunker Hill and Gulf*, Files of Paul Whelan, Plaintiff's attorney.
201. Bill Richards, "Learning Disabilities Raise New Fears of Idaho Lead Poisoning," *Los Angeles Times*, October 21, 1979, 3.
202. Curry, "Children Feared Harmed," 1.
203. Eric G. Comstock, "Medical Toxicology Consultation, Patient 1801," February 4, 1980, *Yoss et al. v. Bunker Hill and Gulf.*
204. Cutchins, "Deposition," July 27, 1981, 84–85.
205. J. William Flynt, Jr., MD, deputy director, Cancer and Birth Defects Division, Bureau of Epidemiology, to the Record, "Lead Poisoning Investigations in Kellogg, Idaho," September 24, 1974, *Yoss et al. v. Bunker Hill and Gulf.*
206. "Statement of Gloria Dixon [pseudonym]," Oversight—Clean Air Act Amendments of 1977, 80-H501-65.
207. Andrea Cornish [pseudonym] to Ed Gallagher, "Letter," March 4, 1980, *Yoss et al. v. Bunker Hill and Gulf*, Files of Paul Whelan, Plaintiff's attorney.
208. Shelly Long, "Summary of the Deposition of John A. Mather, MD, Deposition Taken April 26, 1978," *Yoss et al. v. Bunker Hill and Gulf*, Files of Paul Whelan, Plaintiff's attorney.
209. John A. Mather, "News Release," May 6, 1975, *Yoss et al. v. Bunker Hill and Gulf*, Files of Paul Whelan, Plaintiff's attorney.

210. Cassandra Tate, "Kellogg Warned of Food Danger," *Lewiston Tribune*, May 7, 1975.
211. Ian von Lindern and Tony Yankel to Lee Stokes, "Memorandum: Meeting with CDC and Shoshone Project regarding August, 1975 Survey and September Shoshone Project Committee presentation," July 16, 1975, *Yoss et al. v. Bunker Hill and Gulf*, Files of Paul Whelan, Plaintiff's attorney.
212. Ibid.
213. See, for example, unknown author, "Notes to File, Re: Taylor [pseudonym] v. Bunker Hill," which recounts the public meeting held by the Health Department at the Kellogg Union Hall on October 2, 1974. A parent concerned about her child's blood lead level was advised to see a local physician. The notes read: "A lady who lived in the prime exposed area asked if her child had a level from 54 to 60 whether she should go back to the family physician. This question had two possible answers one of which was to go to anybody else but Bax said go back to the local physician. The lady said that her child with the level of 60 went to bed too early and seems lethargic."
214. Bob Mims, "The Question That Won't Go Away," *Lewiston Tribune*, April 8, 1981, A1. The name Bob Dunn is a pseudonym.
215. "Gulf Resources Unit Settles With 37 More In Lead-Poisoning Case," *Wall Street Journal*, April 21, 1983, 41; and "Children Win Millions in Industrial Pollution Settlement," *American Bar Association Journal*, December 1981, 1604.
216. "Living in Bunker's Shadow," *Spokesman Chronicle*, January 9, 1984, 12.
217. Judy Mills, "Lead Pollution: The Bottom Line Can't Be Figured," *Spokesman Review*, January 8, 1984, E5.
218. Personal communication with Paul Whelan, July 16, 2010, Notes in the author's possession.
219. "Gulf Resources Decides to Close Idaho Mine Unit," *Wall Street Journal*, August 26, 1981, 4.

CHAPTER 5 TACOMA: A DISASTER IS DISCOVERED

Some material presented in this chapter was originally published as Marianne Sullivan, "Contested Science and Exposed Workers," Public Health Reports 122 (July–August 2007): 541–547.

1. William Rodgers to Gordon Johnston, Mayor of Tacoma and Chairman Puget Sound Air Pollution Control Board, May 2, 1972, Environmental Protection Agency Site File, Commencement Bay/Nearshore Tideflats ASARCO Smelter Facility Site File (hereafter ASDSF), 1.1.1.
2. Arthur Dammkoehler, interview, July 20, 2005, Mercer Island, Washington. Digital audio recording. File in author's possession.
3. See for example, R. M. Statnick, *Measurement of Sulfur Dioxide, Particulate, and Trace Elements in Copper Smelter Converter and Roaster/Reverberatory Gas Streams* (October 1974), US EPA, National Environmental Research Center, Research Triangle Park, NC, EPA-650/2-74-111; and Puget Sound Air Pollution Control Agency (PSAPCA), "Environmental Impact Statement for ASARCO, Incorporated. Variance from PSAPCA Regulation I Sections 9.03(b), 9.07(b), and 9.07(c), Final" (September 1981), VI-8.
4. Samuel Milham Jr., MD, and Terrance Strong to Wallace Lane, MD, Washington State Health Services Division, "Memorandum: Tacoma Smelter Study," November 2, 1972, ASDSF 1.1.1; and PSAPCA, "Environmental Impact Statement for ASARCO," VI-8.
5. See, for example, Armand Labbe, "Presentation before the Board of Directors of the Puget Sound Air Pollution Control Agency at the Public Hearing for Proposed Amendments to Regulation I for Controlling the Emission of Arsenic," February 14, 1973,

ASDSF 13.5; Bob Boxberger, "Doctors Discount Danger from Asarco's Arsenic," *Tacoma News Tribune*, October 22, 1975, D10; Herbert G. Lawson, "Dilemma in Tacoma: Smelter's Emissions Threaten Populace, But So Does Possible Loss of 1,000 Jobs," *Wall Street Journal*, July 16, 1975, 36; and "Smelter Promises to Keep Trying," *Vashon Maury Island Beachcomber*, January 22, 1976.

6. Steve Lerner, *Sacrifice Zones: The Front Lines of Toxic Chemical Exposure in the United States* (Cambridge, MA: MIT Press, 2010).

7. James C. Whorton, *Before Silent Spring: Pesticides and Public Health in Pre-DDT America* (Princeton, NJ: Princeton University Press, 1974); James C. Whorton, *The Arsenic Century: How Victorian Britain Was Poisoned at Home, Work, and Play* (New York: Oxford University Press, 2010); and Mark Parascandola, *King of Poisons: A History of Arsenic* (Dulles, VA: Potomac Books, 2012).

8. C. C. Franseen and G. W. Taylor, "Arsenical Keratoses and Carcinomas," *American Journal of Cancer* 22 (1934): 287–307.

9. J. M. Pearce, "Sir Jonathan Hutchinson (1828–1913) and an Early Description of Temporal Arteritis," *Journal of Neurological and Neurosurgical Psychiatry* 57 (1994): 216; and J. Paget, H. H. Bantock, and M. M. De Bartolome, "Reports of Societies: Arsenic Cancer," *British Medical Journal* 2 (1887): 1280–1283.

10. See, for example, U. J. Wile, "Arsenical Cancer with Report of a Case," *Journal of Cutaneous Diseases* (April 1912): 192–199; R. J. Pye-Smith, "Arsenic Cancer, with Description of a Case," *Proceedings of the Royal Society of Medicine* 6 (1913): 229–236; Franseen and Taylor, "Arsenical Keratoses and Carcinomas," 287–307; and S. Ayres and N. P. Anderson, "Cutaneous Manifestations of Arsenic Poisoning," *Archives of Dermatology and Syphilology* 30 (1934): 33–43.

11. See Ayres and Anderson, "Cutaneous Manifestations of Arsenic Poisoning," 33–43, including the discussion by Dr. Jacob C. Geiger, Director of Public Health, San Francisco, who notes the widespread use of arsenic in industry and agriculture; and Franseen and Taylor, "Arsenical Keratoses and Carcinomas," 287–307.

12. T. M. Legge, "Occupational Intoxications: Arsenic Poisoning," in *Diseases of Occupation and Vocational Hygiene*, ed. G. M. Kober and W. C. Hanson (Philadelphia: P. Blakiston's Son & Co., 1916), 5.

13. Alice Hamilton, *Industrial Poisons in the United States* (New York: Macmillan, 1925).

14. Wilhelm C. Hueper, *Occupational Tumors and Allied Diseases* (Springfield, IL: Charles C. Thomas, 1942), 37–38.

15. Ibid., 38.

16. On scrotal cancer, see John Ayrton Paris, *Pharmacologia: Comprehending the Art of Prescribing upon Fixed and Scientific Principles, Together with the History of Medicinal Substances*, vol. 2 (New York: F&R Lockwood, 1823). On bladder cancer, see C. D. Haagensen, "Occupational Neoplastic Disease," *American Journal of Cancer* 15 (1931): 641–703, at 656; and O. Neubauer, "Arsenical Cancer: A Review," *British Journal of Cancer* 1 (1947): 192–251. On lung cancer, see E. Lorenz, "Radioactivity and Lung Cancer; A Critical Review of Lung Cancer in the Mines of Schneeberg and Jaochimsthal," *Journal of the National Cancer Institute* 5 (1944): 1–15. On British studies of the arsenic connection, see A. B. Hill and E. L. Faning, "Studies in the Incidence of Cancer in a Factory Handling Inorganic Compounds of Arsenic: Part I. Mortality Experience in the Factory," *British Journal of Industrial Medicine* 5 (1948): 1–6.

17. Arsenic was a contaminant of aniline dyes. At the E. I. du Pont de Nemours Company (DuPont), arsenic acid was used as an "oxidizing agent" in aniline dye production and could contaminate the finished product. Sometime in the early 1930s DuPont

apparently stopped using arsenic in aniline dye. This information is included as an interesting piece of supportive evidence to a case report of dermatitis in a twenty-three-year-old woman that was attributed to arsenic exposure from the black dye found in a new wool suit. See M. J. Reuter, "The Arsenic Problem," *Archives of Dermatology* 31 (1935): 811–818.
18. Neubauer, "Arsenical Cancer."
19. E. Lynge, "From Cross-sectional Survey to Cohort Study," *Occupational and Environmental Medicine* 66 (2009): 428–429; Hill and Faning, "Studies in the Incidence of Cancer"; F. Roth, "The Sequelae of Chronic Arsenic Poisoning in Moselle Vintners," *German Medical Monthly* 2 (1957): 172–175; and H. S. Osburn, "Cancer of the Lung in Gwanda," *Central African Journal of Medicine* 3 (1957): 215–223.
20. L. S. Snegireff and O. M. Lombard, "Arsenic and Cancer: Observations in the Metallurgical Industry," *AMA Archives of Industrial Hygiene and Occupational Medicine* 4 (1951): 199–205.
21. The Delaney Clause of the Federal Food, Drug, and Cosmetic Act, 21 USC Sec.408 and 409, as amended by the Food Additives Amendment of 1958.
22. Y. E. Lebedeff, ASARCO Central Research Laboratories, to Dr. A. J. Phillips, "Internal ASARCO Memorandum: Arsenic-Cancer-NACA," October 27, 1959, *Branin v. ASARCO*, 93-CV-5132, Accession # 021-04-0102, Box 1-9, Location # 3050865, settlement papers obtained at Seattle Federal Records Center (hereafter *Branin v. ASARCO*).
23. Y. E. Lebedeff to S. S. Pinto, October 28, 1959, *Branin v. ASARCO*; Lebedeff to Phillips, "Internal ASARCO Memorandum," October 27, 1959, *Branin v. ASARCO*.
24. S. S. Pinto and B. M. Bennett, "Effect of Arsenic Trioxide Exposure on Mortality," *Archives of Environmental Health* 7 (November 1963): 583–91, at 590.
25. See Neubauer, "Arsenical Cancer," 192–251.
26. B. L. Vallee, D. D. Ulmer, and W.E.C. Wacker, "Arsenic Toxicology and Biochemistry," *AMA Archives of Industrial Health* 21 (1960): 56–75. Vallee had a long career at Harvard and in the mid-1970s was credited with brokering a controversial deal between Monsanto and Harvard University for cancer research funding. In a 1977 article describing the origin and politics of the deal, the reporter notes that "for years, Vallee has been a corporate consultant to Monsanto." Barbara J. Culliton, "Harvard and Monsanto: The $23 Million Alliance," *Science* 195 (February 25, 1977): 759–763.
27. A. O. Robson and A. M. Jelliffe, "Medicinal Arsenic Poisoning and Lung Cancer," *British Medical Journal* 2 (July 27, 1963): 207–209; J. Wagoner, R. Miller, F. Lundin, J. Fraumeni, and M. Haij, "Unusual Cancer Mortality among a Group of Underground Metal Miners," *New England Journal of Medicine* 269 (1963): 284–289; A. M. Lee and J. F. Fraumeni Jr., "Arsenic and Respiratory Cancer in Man: An Occupational Study," *Journal of the National Cancer Institute* 42 (1969): 1045–1052; and W. P. Tseng, H. M. Chu, S. W. How, J. M. Fong, C. S. Lin, and Shu Yeh, "Prevalence of Skin Cancer in an Endemic Area of Chronic Arsenicism in Taiwan," *Journal of the National Cancer Institute* 40 (1968): 453–463.
28. Whorton, *Arsenic Century*.
29. In Tacoma, when children's urinary arsenic concentrations were studied in the 1970s and 1980s, researchers typically asked whether children had consumed seafood in the days prior to testing, and the results were reported separately. This was not always done, however, and is a consideration in interpreting some of the results from Tacoma.
30. See C. N. Myers and Leon H. Cornwall, "Normal Arsenic and Its Significance from the Point of View of Legal Medicine," *American Journal of Syphilis* 9 (October 1925): 647.

31. S. H. Webster, "The Lead and Arsenic Content of Urines from 46 Persons with No Known Exposure to Lead or Arsenic," *Public Health Reports*, October 3, 1941, 1953–1961.
32. Paul A. Neal, Waldemar C. Dreessen, Thomas I. Edwards, Stewart H. Webster, Harold T. Castberg, and Lawrence T. Fairhall, "A Study of the Effect of Lead Arsenate Exposure on Orchardists and Consumers of Sprayed Fruit," *Public Health Bulletin*, no. 267 (1941).
33. For an in-depth consideration of the USPHS Wenatchee study, see Christopher C. Sellers, *Hazards of the Job: From Industrial Disease to Environmental Health Science* (Chapel Hill: University of North Carolina Press, 1997), 209–214.
34. For example, when a letter to the editor in *JAMA* in 1942 asked, "What is the threshold of urinary arsenic that 'constitutes a danger signal'?," the editors responded that arsenic concentrations up to 700 µg/L could be "regarded as within normal limits" based on the experience with consumers of fruit and orchardists exposed to lead arsenate without apparent ill effects. "Relation of Excretion of Arsenic in Urine to Arsenic Poisoning," *Journal of the American Medical Association* 119 (1942): 854.
35. S. S. Pinto and C. M. McGill, "Arsenic Trioxide Exposure in Industry," *Industrial Medicine and Surgery* 22 (July 1953): 281–287.
36. H. H. Schrenk and Lee Schreibeis, "Urinary Arsenic Levels as an Index of Industrial Exposure," *American Industrial Hygiene Journal* 19 (1958): 225–228.
37. See D. J. Birmingham, M. M. Key, D. A. Holaday, and V. B. Perone, "An Outbreak of Arsenical Dermatoses in a Mining Community," *Archives of Dermatology* 91 (May 1965): 457–464. This article reports on a USPHS investigation in 1965 into an outbreak of arsenical dermatoses in children living in a "camp community" near a gold mine and metal smelter in Nevada. The smelter's dust collection system was not operating efficiently, and some portion of the forty tons of arsenic produced each day was emitted into the air. Arsenic dust covered the ground in visible quantities in the area where workers and their families were living. All of the cats and dogs except one had died, and thirty-two of forty children had dermatoses attributable to arsenic. Despite the dermatologic manifestations of arsenic exposure, the USPHS discounted the possibility that the children's health was being harmed because their urinary arsenic concentrations "compared favorably with the daily average arsenic excretion of 0.82 mg/liter (820 µg/L) that the American Smelting and Refining Company has shown to be benign on the basis of 20 to 30 years of experience." The dermatoses were attributed to local irritation from contact with arsenic, rather than systemic poisoning. Another interesting aspect of this report is that mining and smelting began at this site began in late 1962, and the problem was originally investigated a few months later by the State Department of Health when children began to develop skin lesions. State public health officials tested urinary arsenic levels of children living in the camp and, according to Birmingham et al., "it was ascertained that only trace amounts were being excreted but the dermatoses persisted." State public health officials may have considered children's urinary arsenic levels to be "trace" because of the confusion in the literature over "normal" urinary arsenic levels. The children were not moved out of the area, and another investigation was launched, which is the primary focus of the earlier referenced article. The researchers concluded that "urinary analyses for arsenic, performed on specimens from the school children and the mill workers, indicated that only one individual, a mill worker, was excreting arsenic in excessive amounts.... It is unlikely that the dermatoses observed were manifestations of systemic intoxication."
38. A. Hamilton and H. L. Hardy, *Industrial Toxicology*, 3rd ed. (Littletown, MA: Publishing Sciences Group, 1974), 35.

39. ATSDR, "ToxGuide for Arsenic," October 2007, http://www.atsdr.cdc.gov/toxguides/toxguide-2.pdf.
40. P. J. Landrigan, "Arsenic—State of the Art," *American Journal of Industrial Medicine* 2 (1981), 5–14.
41. Kathleen L. Caldwell, Robert L. Jones, Carl P. Verdon, Jeffery M. Jarrett, Samuel P Caudill, and John D. Osterloh, "Levels of Urinary Total and Speciated Arsenic in the US Population: National Health and Nutrition Examination Survey 2003–2004," *Journal of Exposure Science & Environmental Epidemiology* 19 (2009): 59–68.
42. John Finklea, MD, Division of Health Effects Research, to G. Love, Pollutant Burden Indicator Team, "EPA Memo: Health Hazards of ASARCO Tacoma Smelter," May 1, 1972, ASDSF 1.1.1.
43. Milham and Strong to Lane, "Memorandum," November 2, 1972.
44. PSAPCA, "Environmental Impact Statement for ASARCO," VI-8.
45. Milham and Strong to Lane, "Memorandum," November 2, 1972.
46. "Tacoma Arsenic Report," *Seattle Post-Intelligencer*, November 16, 1972, A17; and Malcolm MacNey, "Scientist Says High Arsenic Level Comes from Air, Ground," *Tacoma News Tribune*, November 17, 1972, 2.
47. "Tacoma Arsenic Report."
48. William Barthel, chief of the toxicology section, Department of Health, Education and Welfare, to Terry Strong, State of Washington Department of Social and Health Services, "Letter with attached table indicating blood lead, hair arsenic and urinary arsenic test results for Fern Hill and Ruston school children," November 16, 1972, ASDSF 1.6.6.2; and Samuel Milham Jr., MD, to Armand Labbe, manager, ASARCO Tacoma, January 8, 1973, ASDSF 1.1.1.
49. Terrance R. Strong, section head radiation, Chemical and Physical Hazards Section, to Sam Reed, chief of the Office of Environmental Health Programs, State of Washington Department of Social and Health Services, "Memorandum: Summary of the Smelter Study," September 18, 1974, ASDSF 1.1.1. Although State Health Department officials dismissed the possibility of a lead problem near the smelter, John Roberts, an engineer with PSAPCA, remained concerned about lead exposure in the community. He identified what he thought was an unusual clustering of twelve children classified by the Tacoma school district as "mentally retarded" who lived close to the smelter. Thinking that lead exposure may have contributed to these children's cognitive deficits, Roberts pushed for further study. In consultation with Landrigan and Milham, plans for further testing were developed but apparently not implemented. See Philip Landrigan, MD, to John Roberts, PSAPCA, February 13, 1974, ASDSF 1.1.1; and John Roberts to chief of engineering, "PSAPCA Memorandum: Meeting with Drs. Philip Landrigan and Sam Milham Regarding Health Studies Near the Tacoma Smelter," May 20, 1974, ASDSF 1.1.1. Investigations of BLLs were much more extensive in other smelting communities. For example, in El Paso, at least 132 one- to four-year-olds and 225 five- to nine-year-olds received blood lead tests. Many more older children and adults were also tested. Additionally, more than four thousand samples of soil and household dust were tested for lead content. See "Human Lead Absorption—Texas," *Morbidity and Mortality Weekly Report* 22 (December 8, 1973): 405–407, http://www.cdc.gov/mmwr/preview/mmwrhtml/lmrk095.htm. Ambient air concentrations of lead were not nearly as high in Tacoma as those reported in El Paso.
50. Webster, "Lead and Arsenic Content of Urines."
51. Pinto and Bennett, "Effect of Arsenic Trioxide Exposure," 583–591. NIOSH questioned ASARCO's categorization of exposed and unexposed workers in its analysis of this

study for the occupational arsenic standard. NIOSH researchers showed that when "exposed" and "unexposed workers" were examined together, the study showed that a higher proportion of deaths in smelter workers were due to respiratory cancer compared to Washington State as a whole. See US Department of Health and Human Services, National Institute of Occupational Safety and Health, *Criteria for a Recommended Standard: Occupational Exposure to Inorganic Arsenic (New Criteria—1975)*, NIOSH Publication No. 75–149 (1975), 40.

52. Pinto and McGill, "Arsenic Trioxide Exposure in Industry," 281–287.
53. Lee and Fraumeni, "Arsenic and Respiratory Cancer in Man," 1045–1052.
54. K. W. Nelson to D. H. Soutar, vice president of industrial relations, "Confidential Memorandum: Cloak and Dagger Department," September 11, 1968, *Branin v. ASARCO*.
55. Wallace Lane, MD, to Arthur Dammkoehler and Board of Directors, PSAPCA, November 9, 1972, ASDSF 1.1.1; Arthur Dammkoehler to Wallace Lane, MD, "Lead and Arsenic Studies Vicinity of ASARCO Copper Smelter," November 13, 1972, ASDSF 1.1.1; and Arthur Dammkoehler to James Agee, administrator, Region 10 EPA, "Request for EPA Technical Assistance in Establishing Ambient Air and Emission Standards for Arsenic," November 14, 1972, Environmental Protection Agency ASARCO Administrative Record on microfilm (hereafter ASAR, with microfilm number and start frame number following), ASA208 Frame 1375.
56. Simon Strauss to Unknown Recipients, "Memorandum," January 4, 1972, *Branin v. ASARCO*.
57. Editorial, "The New ASARCO," *Tacoma News Tribune*, July 19, 1972, D2.
58. Armand Labbe, Tacoma Smelter manager, to L. C. Travis, ASARCO Salt Lake City Office, "Memorandum, with Soil Sampling Results Attached," May 9, 1972, ASDSF 1.6.1.
59. Lee Travis to Kenneth Nelson, director, Department of Environmental Sciences, "Memorandum: Tacoma Soil Samples," May 15, 1972, *Branin v. ASARCO*.
60. Labbe to Travis, "Memorandum, with Soil Sampling Results Attached."
61. C. H. Hine, MD, PhD, consultant in occupational medicine and industrial toxicology, to Kenneth W. Nelson, "Notes from Discussion with Dr. Sherman Pinto," October 25, 1972, *Branin v. ASARCO*.
62. Kenneth Nelson to Armand Labbe, "Draft Memorandum of Understanding" (ASARCO draft) attached to "Memorandum: Arsenic," December 18, 1972, ASDSF 1.1.1.
63. Testimony of Dr. Ted A. Loomis, attachment to C. J. Newlands to R. J. Muth, February 16, 1973, ASDSF 1.6.3.
64. "Summary of Testimony: American Smelting and Refining Company, OSHA Fact Finding Hearing on Inorganic Arsenic," September 20, 1974, *Branin v. ASARCO*.
65. See, for example, Labbe, "Presentation before the Board of Directors," February 14, 1973, ASDSF 13.5; Lawson, "Dilemma in Tacoma"; and "Smelter Promises to Keep Trying."
66. Paul Heilman, "Arsenic Content of Soils in the Vicinity of the Ruston Smelter," Western Washington Research and Extension Center (n.d., published in early 1970s), ASDSF 1.6.1; H. Alsid, B. Amundson, G. Hofer, and D. Lutrick, "The Tacoma Air Pollution Study (July–August 1970)," September 30, 1970, Air & Resources Program, Water and Air Resources Division, Department of Civil Engineering, University of Washington, Seattle, University of Washington Libraries, Special Collections, University Archives, American Lung Association of Washington Records (hereafter ALA 5271–1 B4 F15); John Roberts, air pollution engineer, to chief of engineering, "Memorandum: Preliminary Analysis of the Content of Vegetation near ASARCO Stack," May 9, 1973, ASDSF 1.6.2; Dr. James Everts to James Agee, "Memorandum: Mercury Lead and Arsenic

Contamination in Cabbage and Lettuce," November 21, 1973, ASDSF 1.6.2; Michael Bothner, research associate, University of Washington, to John Roberts, October 25, 1973, ASDSF 1.6.1; Eric A. Crecelius and D. Z. Piper, "Particulate Lead Contamination Recorded in Sedimentary Cores from Lake Washington, Seattle," *Environmental Science and Technology* 7 (November, 1973): 1053–1055; Eric A. Crecelius, "The Geochemistry of Arsenic and Antimony in Puget Sound and Lake Washington" (PhD diss., University of Washington, July 19, 1974); and Eric A. Crecelius, C. J. Johnson, and G. C. Hofer, "Contamination of Soils Near a Copper Smelter by Arsenic, Antimony and Lead," *Water, Air and Soil Pollution* 3 (1974): 337–342.

67. Puget Sound Air Pollution Control Agency, "Staff Report: Section 9.19, Regulation I Arsenic Emission Standard," February 14, 1973, ASDSF 1.1.1; and US Environmental Protection Agency, Strategies and Air Standards Division, Office of Air Quality Planning and Standards, "Air Pollutant Assessment Report on Arsenic," July 1976, ASAR ASA212 Frame 0903.

68. Statnick, *Measurement of Sulfur Dioxide*. Per-year emissions estimates assume round-the-clock operation.

69. Puget Sound Air Pollution Control Agency, *The Concentration of Lead, Arsenic, Mercury, Cadmium, Sulfur Dioxide and Suspended Sulfates Downwind from the Tacoma Smelter, the Impact and Control Status, and Benefits from Reduction, Interim Report for Presentation to the ASARCO Impact Task Force Being Coordinated by EPA*, April 22, 1974, ASDSF 1.6.3; and Statnick, *Measurement of Sulfur Dioxide*.

70. Environmental Protection Agency, "Background Document on American Smelting & Refining Company (ASARCO), Tacoma Heavy Metals" (n.d., 1974 or 1975), ASDSF 1.2.

71. Hilman Ratsch, *Heavy-Metal Accumulation in Soil and Vegetation from Smelter Emissions* (August 1974), US EPA National Environmental Research Center, Office of Research and Development, EPA-660/3–74–012, 2.

CHAPTER 6 A CARCINOGENIC THREAT

Some material presented in this chapter was originally published as Marianne Sullivan, "Contested Science and Exposed Workers," *Public Health Reports* 122 (July–August 2007): 541–547.

1. Thank you to Dr. Michael Edelstein of Ramapo College of New Jersey for pointing out the relationship between restaurant choice and municipal allegiance on this particular block.

2. 2010 U.S. Census.

3. B. Halman and J. Cammon Findlay, *Vashon-Maury Island* (Charleston, SC: Arcadia Publishing, 2011).

4. Dr. Carl J. Johnson, MD, MPH, to Dr. Lawrence Bergner, director of public health, Seattle–King County Health Department, "Interim Report Re: Evaluation of Environmental Contamination of Vashon and Maury Islands by Lead and Arsenic Emitted by the Tacoma Smelter," n.d., King County Archives (hereafter KCA), 112 Box Seattle King County Health Department Director Issue Files 1965–1980, Folder: Vashon-Maury 1973–74.

5. Vashon cows had average hair arsenic levels of 8,900 ppb, compared to 460 ppb found in cows thirty-five miles away. Blood arsenic was approximately three times as high, and arsenic in the cows' milk was about twice as high. Lead and arsenic in drinking water in two reservoirs, however, met federal standards; see R. M. Orhiem, L. Lippman, C. J. Johnson, H. H. Bovee, "Lead and Arsenic Levels of Dairy Cattle in

Proximity to a Copper Smelter," *Environmental Letters* 7, no. 3 (1974): 229–236. Seventeen soil samples from the island showed much higher than background levels of arsenic in soil. The highest soil arsenic sample was 120 ppm from the southern end of Vashon; see G. C. Hofer, M. J. Svoboda, A. E. Parlier, and R. D. Pollock, Puget Sound Air Pollution Control Agency, "Interim Report: Lead and Arsenic Content in the Soil Affected by the Copper Smelter located in Tacoma, Washington," October, 1972, KCA Box Seattle King County Health Department Director Issue Files 68–79, Folder 10 Lead Poisoning 1972–79. Testing of children on southern Vashon Island was planned for winter, when prevailing winds blew smelter pollutants over the island. In January of 1973 thirty children who lived at various locations on the island were tested. The children's hair arsenic levels ranged from 0.13 ppm to 21.1 ppm. Normal is considered less than 1.0 ppm. Urinary arsenic concentrations in five children were above 20 µg/L and the highest was 220 µg/L. Five children had blood lead levels (BLL) above 40 µg/dL, with the highest BLL at 55 µg/dL. The BLLs were not considered to be high enough to be of "immediate concern"; see C. J. Johnson and L. Lippman, "Environmental Contamination with Lead and Arsenic from a Copper Smelter," Paper presented at the Pacific Northwest International Section of the Air Pollution Control Association in Seattle, Washington, No. 73-AP-37, November 29, 1973. Dr. Lawrence Bergner thought that exposure levels measured on Vashon might not reflect typical winter exposure. Urinary arsenic concentrations are indicative of exposure in the preceding day or two, and in the week before the testing was done the winds were predominantly out of the north, resulting in less exposure to smelter pollution; see Lawrence Bergner, MD, MPH, director of public health, Seattle–King County Department of Public Health to Arthur Dammkoehler, February 13, 1973, Environmental Protection Agency ASARCO Administrative Record, microfilm (hereafter ASAR, with microfilm number and start frame number following), ASA214 Frame 2225.

6. Hofer, Svoboda, Parlier, and Pollock, "Interim Report," October 1972; Samuel Milham and Terrance Strong, "Human Arsenic Exposure in Relation to a Copper Smelter," *Environmental Research* 7 (1974): 176–182; and Johnson and Lippman, "Environmental Contamination with Lead and Arsenic."

7. Gregory L. Glass, *Credible Evidence Report: The ASARCO Tacoma Smelter and Regional Soil Contamination in Puget Sound, Final Report*, September 2003, 24.

8. K. W. Olden and J. Guthrie, "Air Toxics Regulatory Issues Facing Urban Settings," *Environmental Health Perspectives* 104, Suppl. 5 (1996): 857–860.

9. On local concerns, see Robert Burd, director Air and Water Programs Division, to Arthur Dammkoehler, December 7, 1972, ASAR, ASA211 Frame 1347; Terrance R. Strong to Smelter Task Force, "Meeting Minutes: Smelter Task Force Meeting," January 3, 1975, ASAR ASA211 Frame 1437; and William Lappenbusch, chief of Radiation, Toxic Substances and Noise Section, to Douglas Hansen, director Air and Hazardous Materials Division, "Internal EPA Memorandum: Heavy Metal Emissions—ASARCO," January 7, 1975, Environmental Protection Agency Site File, Commencement Bay/Nearshore Tideflats ASARCO Administrative Record (hereafter AR), 11.3.1. On funding for health studies, see James Everts, research and development representative, Region 10, Requestor, "Environmental Research Needs: Hazard Evaluation of the Tacoma, Washington, ASARCO Copper Smelter," January 31, 1974, AR 1.1; Douglas Hansen, director Air and Hazardous Materials Division, to Clifford Smith Jr., regional administrator, "Internal EPA Memorandum: ASARCO Strategy Document," February 26, 1975, Environmental Protection Agency Site File, Commencement Bay/Nearshore Tideflats ASARCO Smelter Facility Site File (hereafter ASDSF), 1.1.1; and James B. Weigold, Office of Air Quality Planning and

Standards to Distribution, "Meeting Summary: 11/20/75 Meeting with Doug Hansen on ASARCO-Tacoma Smelter," November 24, 1975, ASAR ASA208 Frame 1715.
10. Jeff Weathersby, "EPA Rejected Arsenic Standards as Early as '76," *News Tribune*, October 10, 1983, B1.
11. B. J. Steigerwald, director, Office of Air Quality Planning and Standards, to Roger Strelow, assistant administrator for Air and Waste Management, "Memorandum: Arsenic," February 1975, ASDSF 1.1.1.
12. Department of Health and Human Services (US), National Institute of Occupational Safety and Health, *Criteria for a Recommended Standard: Occupational Exposure to Inorganic Arsenic (New Criteria—1975)*, (1975), HEW Publication No. (NIOSH) 75–149, quote at 89. See also H. P. Blejer and W. Wagner, "Inorganic Arsenic—Ambient Level Approach to the Control of Occupational Cancerigenic Exposures," *Annals of the New York Academy of Sciences* 271 (1976): 179–186.
13. J. A. Newman, V. E. Archer, G. Saccomanno, M. Kuschner et al., "Histologic Types of Bronchogenic Carcinoma among Members of Copper-Mining and Smelting Communities," *Annals of the New York Academy of Sciences* 271 (1976): 260–268; and W. J. Blot and J. F. Fraumeni Jr., "Arsenical Air Pollution and Lung Cancer," *Lancet*, July 26, 1975, 142–144.
14. Arsenic, Inorganic, 40 Fed. Reg. 14 (Jan. 21, 1975), 3392–404. The final OSHA standard for arsenic in workplace air was set at 10 micrograms per cubic meter averaged over 8 hours; Toxic and Hazardous Substances, 29 CFR 1910.1018, http://www.osha.gov/pls/oshaweb/owadisp.show_document?p_table=STANDARDS&p_id=10023.
15. Steigerwald to Strelow, "Memorandum: Arsenic."
16. Ibid.
17. See, for example, "Statement of Kenneth W. Nelson before the United States Department of Labor Occupational Safety and Health Administration, In Re Proposed Standard for Occupational Exposure to Inorganic Arsenic," April 16, 1975, Docket Number H037A Exhibit 29(p). Obtained from the Occupational Safety and Health Administration.
18. Milham and Strong, "Human Arsenic Exposure."
19. Armand Labbe, manager, Tacoma Smelter, to Fellow ASARCO Employee, March 13, 1974, *Branin v. ASARCO*, 93-CV-5132, Accession # 021-04-0102, Box 1-9, Location #3050865, settlement papers obtained at Seattle Federal Records Center (hereafter *Branin v. ASARCO*).
20. Anne Fischel and Lin Nelson, "Their Mines Our Stories, Ruston, WA," Interview with Rodger Jones, http://www.theirminesourstories.org/?cat=4, accessed April 12, 2012. See, for example, "ASARCO Fears Shutdown," *Tacoma News Tribune*, January 27, 1976, A1; Margaret Ainscough, "Asarco Letter to Employees Blasted," *Tacoma News Tribune*, April 14, 1977, B1; Roger Ainsley, "How Safe Should the Smelter Be?" *Seattle Post Intelligencer, Northwest Magazine*, March 9, 1980, 5; Bob Lane, "Smelter Firm Again Threatens Shutdown," *Seattle Times*, May 4, 1971, C15; Jack Pyle, "Smelter Needs Aid, Says Federal Official," *Tacoma News Tribune*, April 8, 1977, A1; and "Tacoma Smelter Threatens to Close," *Seattle Post-Intelligencer*, August 6, 1983, C1.
21. Steigerwald to Strelow, "Memorandum: Arsenic."
22. J. Rabovsky, "Are Separate Standards for Occupational and Environmental Exposures Good Public Health Policy?" *New Solutions* 15 (2005): 211–219.
23. Steigerwald to Strelow, "Memorandum: Arsenic."
24. Puget Sound Air Pollution Control Agency, *Air Contaminant Emissions from ASARCO's Tacoma Smelter, Chronology of Control Efforts*, n.d. Obtained from Puget Sound Clean Air Agency.

25. Steigerwald to Strelow, "Memorandum: Arsenic"; and Committee on Biologic Effects of Environmental Pollutants, *Arsenic: Medical and Biological Effects of Environmental Pollutants* (Washington, DC: National Academy of Sciences Press, 1977).
26. Janet Chalupnik, interview, June 17, 2005, Edmonds, Washington. Digital audio recording. File in author's possession; and "Before the Pollution Control Hearings Board of the State of Washington, In the Matter of the Application for Variance of the American Smelting and Refining Company to the Puget Sound Air Pollution Control Agency, Notice of Cross-Appeal No. HB-70–38," n.d., University of Washington Libraries, Special Collections, University Archives, American Lung Association of Washington Records (hereafter ALA) 5271, Box 4 Folder 4.
27. Arthur Dammkoehler, interview, July 20, 2005, Mercer Island, Washington. Digital audio recording. File in author's possession.
28. Puget Sound Air Pollution Control Agency, "Environmental Impact Statement for ASARCO, Incorporated. Variance from PSAPCA Regulation I Sections 9.03(b), 9.07(b), and 9.07(c), Final," September 1981, III-2, Table III-1.
29. Samuel Milham Jr., MD, to Arthur Dammkoehler, October 1, 1976, ALA 5271-001 B3 F56. Three tables with urinary arsenic concentrations attached.
30. K. Wicklund and L. Harter, "Unpublished manuscript: Urinary Arsenic Levels of Residents Living Near the ASARCO Smelter, Tacoma from 1972–1983," n.d., Washington State Department of Ecology, ASARCO electronic files (hereafter Ecology electronic files); Puget Sound Air Pollution Control Agency, "Environmental Impact Statement for ASARCO," VI-7–9.
31. Compare results from the June 3, 1975, sampling. In thirty-six samples from Ruston schoolchildren without prior seafood consumption, the State Department of Health reported urinary arsenic concentrations ranging from 20 to 660 µg/L, with a mean of 87 µg/L, see ibid., VI-8. ASARCO's laboratory reported a mean urinary arsenic concentration of <60 µg/L, based on thirty-eight samples, which included those with seafood consumption and corrected for specific gravity. In this sample, ASARCO researchers apparently dropped the highest value of 395 µg /L, considering it an anomaly, although the child's father worked at the smelter and the family lived one block from the stack. M. O. Varner to Armand Labbe, "Confidential Memorandum: Schoolchildren Urinary Arsenic Values," June 19, 1975, ASDSF 1.6.6.1.
32. Curtis E. Dungey, plant industrial hygienist, to Mr. & Mrs. Thomas X [name withheld], Tacoma, WA, October 26, 1977, ASDSF 1.6.6.1.
33. Curtis E. Dungey to Parents, "Letter to Multiple Ruston/Tacoma Parents," October 16, 1975, *Branin v. ASARCO*.
34. Office of Environmental Health Programs Occupational Health Section, Washington State Department of Social and Health Services, "Human Arsenic Exposure in Relation to a Copper Smelter," n.d., ASDSF 13.4.
35. The CDC/EPA national smelter study was unpublished at this time and likely would not have been available to Milham. See E. L. Baker, C. G. Hayes, P. J. Landrigan, J. L. Handke, R. T. Leger, W. J. Housworth, and J. M. Harrington, "A Nationwide Survey of Heavy Metal Absorption in Children Living Near Primary Copper, Lead and Zinc Smelters," *American Journal of Epidemiology* 106 (1977): 261–273.
36. ASARCO News, "Press Release: Tacoma Smelter Arsenic Emissions Pose No Danger to Maury and Vashon Island Residents," July 15, 1976, ASDSF 13.4.
37. Milham to Dammkoehler, October 1, 1976. "Drummond" is a pseudonym.
38. Curtis E. Dungey to Armand Labbe, "Memorandum: Environmental Study of Drummond Home [pseudonym]," November 24, 1976, ASDSF 1.6.3.

39. Donald A. Robbins, ASARCO chief chemist, to Curtis E. Dungey, January 17, 1977, ASDSF 1.6.3.
40. Curtis E. Dungey to M. O. Varner, manager, Department of Environmental Sciences, "Memorandum: Environmental Health Study Drummond House and Nearby Vicinity," February 10, 1977, ASDSF 1.6.3.
41. Wicklund and Harter, "Urinary Arsenic Levels," 5.
42. See Bob Boxberger, "Arsenic-Cancer Link Indicated," *Tacoma News Tribune*, December 19, 1975, B11. The study was subsequently published as S. S. Pinto, P. E. Enterline, V. Henderson, and M. O. Varner, "Mortality Experience in Relation to a Measured Arsenic Trioxide Exposure," *Environmental Health Perspectives* 19 (1977): 127–130.
43. Pinto, Enterline, Henderson, and Varner, ""Mortality Experience."
44. See Robert N. Proctor, *Cancer Wars: How Politics Shapes What We Know and Don't Know About Cancer* (New York: Basic Books, 1995), 161–166.
45. Puget Sound Air Pollution Control Agency, "Transcript of the Public Hearing Relative to ASARCO Variance Application Number 157 Held before the Board of Directors," January 27, 1976, ASAR ASA218 Frame 0279, 53–66.
46. M. O. Varner to W. K. Murray, "Internal ASARCO Memorandum: Paper Entitled, 'Mortality Experience in Relation to a Measured Arsenic Trioxide Exposure, by S. S. Pinto, et al.,'" September 22, 1976, ASDSF 1.6.4.
47. Pinto, Enterline, Henderson, and Varner, "Mortality Experience."
48. Puget Sound Air Pollution Control Agency, "Resolution 359: Resolution of the Board of Directors of Puget Sound Air Pollution Control Agency Granting Unto ASARCO Incorporated, Tacoma Smelter, A Variance 9.03(b), 9.07(b)(c) and 9.19 (c) of Regulation I," February 19, 1976, ALA 5271–1 B4 F18.
49. K. W. Nelson to Armand Labbe, "Memorandum: Dr. DiGiacomo's Proposed Study of Tacoma Children," December 8, 1977, ASDSF 1.6.6.2.
50. Strong to Smelter Task Force, "Meeting Minutes."
51. Assistant administrator for Enforcement and General Counsel and assistant administrator for Research and Development to assistant administrator for Planning and Management, "Internal EPA Memorandum: Funding for a Nationwide Smelter Survey," December 26, 1974, ASDSF 1.1.1.
52. H. P. Blejer, "Arsenic: Occupational Sentinel of Community Disease," Reprint from International Conference on Heavy Metals in the Environment, Toronto, October 27–31, 1975, 317–328.
53. Committee on Biologic Effects of Environmental Pollutants, *Arsenic*, 223.
54. Walter C. Barber, director, Office of Air Quality Planning and Standards, to Honorable Lloyd Meeds, House of Representatives, January 7, 1977, ALA 5271-001 B3 F56.
55. Roy E. Albert, MD, chairman, Carcinogen Assessment Group, to Joseph Padgett, director, Strategies and Air Standards Division, "Memorandum: Carcinogenicity of Arsenic," April 29, 1977, ASAR ASA214 Frame 2237.
56. A. W. Reitze and R. Lowell, "Control of Hazardous Air Pollution," *Boston College Environmental Affairs Law Review* 28 nos. 2/3 (Winter 2001): 229–362.
57. "Group Seeking Limits on Arsenic in the Air Files Suit against U.S.," *Wall Street Journal*, August 16, 1978, 3.
58. U.S. Environmental Protection Agency, "Oral History Interview-1: William D. Ruckelshaus," January 1993. http://www2.epa.gov/aboutepa/william-d-ruckelshaus-oral-history-interview.
59. Peter Behr, "Reagan Team Takes Stock of Prospects for Reshaping Regulation," *Washington Post*, February 6, 1981, A2.

60. David Hoffman and Cass Peterson, "Burford Quits As EPA Administrator," *Washington Post*, March 10, 1983, A1.
61. David Andrews, former EPA employee, quoted in J. A. Mintz, *Enforcement at the EPA: High Stakes and Hard Choices* (Austin: University of Texas Press, 1995), 60.
62. See, for example, Pacific Environmental Services, Inc. to Alfred Vervaert and Graham Fitzsimons, "Memorandum: Minutes of Meeting with ASARCO-Tacoma on 3/16/81, Copper/Arsenic NESHAP Project," March 30, 1981, ASAR ASA210 Frame 2017.
63. Region 10 EPA, "Spill Report: Taken by Joan McNamee," February 4, 1981, ASDSF 1.1.2.
64. Ibid.
65. For example, between March and April of 1982 the station near the main stack recorded concentrations "over 10 $\mu g/m^3$ on three consecutive days." PSAPCA head Dammkoehler reported this to the board with a note of frustration: "The OSHA arsenic standard requires that a respirator be worn in any area inside the plant where the arsenic concentration exceeds 10 $\mu g/m^3$ (8-hour average). ASARCO has been asked to investigate and furnish an explanation as well as measures to prevent a repetition." A. R. Dammkoehler to PSAPCA Board of Directors, "Memorandum: ASARCO Status Report," August 5, 1982, ALA 5271-001 B4 F2.
66. J. Hartley, S. Milham, and P. Enterline, "Lung Cancer Mortality in a Community Surrounding a Copper Smelter," n.d., unpublished manuscript. Washington State Historical Society (hereafter WSHS), Pugnetti papers, Box 5 Folder 14; and L. Polissar, R. K. Severson, Y. T. Lee, "Cancer Incidence in Relation to a Smelter," October 1979, unpublished manuscript, ASDSF 1.6.6.3.
67. Samuel Milham, "Summary: A Long Term Follow-up of School Children Exposed to Atmospheric Arsenic," n.d., unpublished manuscript, WSHS, Pugnetti papers, Box 5 Folder 14.
68. Lew Kittle to Rick Pierce and Craig Baker, "Memorandum: Quartermaster Harbor Fish Kill, February 23, 1981," March 17, 1981, Washington State Archives (hereafter WSA), Ecology Toxics Cleanup Program, Box 26 ASARCO Site, F: ASARCO RI Preliminary Data.
69. "Summary of Wastewater Discharges from ASARCO, Inc., Tacoma Smelter," January 30, 1976, WSA, Ecology, Box 372, Water Quality File 67–86, F: Correspondence 72–77.
70. Richard Pierce, regional engineer, to Norman Glenn, regional manager, "Memorandum: Recommendation of Penalty to ASARCO, Slag Processing Operation, Tacoma," February 15, 1984, WSA, Ecology, Box 25 Toxics Cleanup Program ASARCO Site 62–86.
71. Jeff Weathersby, "Puyallups Threaten Lawsuit to Close Asarco Smelter," *Tacoma News Tribune*, April 30, 1981, A4.
72. Jeff Weathersby, "Decision Delayed on Smelter Bid for Air Pollution Variance," *Tacoma News Tribune*, October 23, 1981, A3.
73. A. R. Dammkoehler, *History of Relations: Puget Sound Air Pollution Control Agency and American Smelting and Refining Company Copper Smelter March, 1968–January, 1971*, Issued by Puget Sound Air Pollution Control Agency, January 22, 1971, Ecology electronic files.
74. R. J. Vong and A. P. Waggoner, "Measurements of the Chemical Composition of Western Washington Rainwater, 1982–1983," July 29, 1983, Report to EPA, Ecology electronic files.
75. Edwin L. Coate, EPA Region 10 deputy administrator, "Speech before the Third Annual Conference of the Northwest Association for Environmental Studies: Acid Rain in the Northwest, University of Victoria, British Columbia," November 2, 1984, National Archives and Records Administration, Pacific Region (hereafter NARA PR), RG 412-87-006.

76. Donald R. Dubois, regional administrator, to Wilbur G. Hallauer, director, Department of Ecology, June 1979, ASAR ASA209 Frame 1574.
77. Puget Sound Air Pollution Control Agency, "Action on ASARCO Copper Smelter Variance Application," November 13, 1981, ASAR ASA210 Frame 0493.
78. J. J. Bromenshenk, S. R. Carlson, J. C. Simpson, and J. M. Thomas, "Pollution Monitoring of Puget Sound with Honey Bees," *Science* 27 (February 1985): 632–634.
79. Ibid.
80. J. J. Bromenshenk to Puget Sound beekeepers, November 22, 1982, Ecology electronic files.
81. Bromenshenk et al., "Pollution Monitoring of Puget Sound."
82. K. Lowry, *Arsenic and Cadmium Levels Found in Garden Soils in Tacoma, Washington*, Tacoma Pierce County Health Department and Seattle-King County Health Department, April 1983, Ecology electronic files; and Jean Snell, president, and Frank Jackson, immediate past president, Vashon Island Community Council to A. R. Dammkoehler and L. Edwin Coate, acting regional administrator Region 10, May 17, 1983, ASAR ASA216 Frame 0856.
83. Arthur R. Dammkoehler to Dr. Ray M. Nicola, "Particulate Matter Fallout from the Tacoma Smelter," July 16, 1981, ASAR ASA216 Frame 0598.
84. Douglas L. Pierce, Tacoma Pierce County Environmental Health Division, to Editor, *Tacoma News Tribune*, August 5, 1981, ASAR ASA216 Frame 0609.
85. John Gillie, "Gardeners Sue Asarco-Ruston Couple Angry over Garden Loss," *Tacoma News Tribune*, May 19, 1983, A3.
86. Ibid.
87. Jeff Weathersby, "12 Taken to Hospital after Smelter Coughs Arsenic Dust on Road," *Tacoma News Tribune*, January 8, 1982, 1.
88. John A. McCarthy, attorney-at-law, to Michael Thorp, attorney-at-law, January 14, 1982, *Branin v. ASARCO*.
89. Don R. Goodwin to Ken Nelson, June 19, 1981, ASAR ASA210 Frame 1651.
90. Bill Tobin, interview, August 5, 2005, Vashon Island, Washington. Digital audio recording. File in author's possession.
91. Ibid.
92. *Bradley v. American Smelting and Refining Company*, 104 Wn.2d 677, 709 P.2d 782.
93. Bill Tobin interview.
94. Ibid.
95. Jeff Weathersby, "Couple's Suit over Arsenic from ASARCO Goes to Trial," *Tacoma News Tribune*, April 8, 1986, B2.
96. "U.S. Must Face Minerals Issues Now: ASARCO's Barber Tells Strategic Resources Conference," *Engineering and Mining Journal*, January 1982, 9.
97. Michael Malone, "The Collapse of Western Metal Mining: An Historical Epitaph," *Pacific Historical Review* 55 (August 1986): 455–464.
98. Joel Connelly, "Tacoma Smelter Threatens to Close," *Seattle Post-Intelligencer*, August 6, 1983, C1.
99. "Response to EPA Questions on ASARCO Tacoma Copper Smelter," n.d., ASAR ASA217 Frame 1091.
100. "Inorganic Arsenic Is Now Listed among Hazardous Air Pollutants," *New York Times*, June 8, 1980, 44; and "Arsenic Emissions Cut of 17% Is Urged by EPA," *Wall Street Journal*, July 13, 1983, 40.
101. U.S. EPA, Office of Air, Noise and Radiation Emission Standards and Engineering Division, "Meeting Minutes: National Air Pollution Control Techniques Advisory

Committee, Minutes of Meeting March 17 and 18, 1981," April 17, 1981, WSA, Ecology Box 372 Water Quality File, F: Correspondence 1978–1981.
102. Gene Lobe, chairman, PSAPCA Board of Directors, to Anne McGill Gorsuch, administrator, EPA, "Request for EPA Promulgation of Emissions Standards for Control of Arsenic," May 8, 1981, ASAR ASA209 Frame 1875.
103. *New York v. Ann Gorsuch*, No. 81 Civ. 6678 (WCC) U.S. District Court for the Southern District of New York; 554 F. Supp. 1060 (January 12, 1983).
104. U.S. EPA, "Oral History Interview-1: William D. Ruckelshaus."
105. "Less-Stringent Rules on Arsenic Pollution Expected to Be Proposed by Chief of EPA," *Wall Street Journal*, July 8, 1983, 2.
106. Environmental Protection Agency, *Executive Summary: National Emission Standards for the Hazardous Air Pollutant (NESHAPS) for Inorganic Arsenic*, n.d., Environmental Protection Agency Site File, Commencement Bay/Nearshore Tideflats Ruston/North Tacoma Site File (hereafter RNTSF), 1.1.2.
107. Goodwin to Nelson, June 19, 1981.
108. EPA, *Executive Summary*.
109. National Emission Standards for Hazardous Air Pollutants; Proposed Standards for Inorganic Arsenic, 48 Fed. Reg. 140 (July 20, 1983), 33123.
110. EPA, *Executive Summary*.
111. John M. Berry, "U.S. Economy Set for Repeated Slumps," *Washington Post*, January 10, 1982, G1; Patrick Boyle, "Profile of Unemployed Changes," *Los Angeles Times*, October 9, 1982, A1; and "Growing Anxiety in Seattle," *New York Times*, June 6, 1982, F23. See also Susan E. Clarke and Gary L. Gaile, *The Work of Cities* (Minneapolis: University of Minnesota Press, 1998), 163–179.
112. Editorial, "Weighing Risks of Arsenic and Pondering the Politics," *Tacoma News Tribune*, July 15, 1983, A15.
113. Editorial, "Tacoma's Role in EPA's Rule," *Seattle Post-Intelligencer*, July 14, 1983, A10.
114. "Mr. Ruckelshaus as Caesar," *New York Times*, July 16, 1983, 22.
115. U.S. Environmental Protection Agency, "Statement of Joseph A. Cannon, Assistant Administrator for Air and Radiation, U.S. EPA, before the Committee on Environment and Public Works, United States Senate," February 24, 1984, NARA PR, RG 412-87-006.
116. W. D. Ruckelshaus, "Science, Risk, and Public Policy," *Science* 221 (September 9, 1983): 1026–1028.
117. E. Randolph, "What Cost a Life? EPA Asks Tacoma," in *Microeconomic Principles in Action*, ed. R. L. Moore and J. D. Whitney, 108–111 (Englewood Cliffs, NJ: Prentice Hall, 1990).
118. Frank W. Jackson and Jeanne Snell to William Ruckelshaus, July 8, 1983, ASAR ASA216 Frame 0752.
119. Larry Lange, "Downwind of ASARCO and Worried," *Seattle Post-Intelligencer*, August 11, 1983, A5; and Environmental Protection Agency, "Meeting Notes: Tacoma Public Meeting, Questions from Group 2, Recorder: Lori Cohen," August 18, 1983, ASAR ASA218 Frame 2091.
120. Wayne E. Grotheer, environmental engineer, EPA, "Meeting Notes: Meeting with Steelworkers Union Local in Ruston, July 30, 1983, Concerning ASARCO-Arsenic NESHAP," August 22, 1983, ASAR ASA209 Frame 0307.
121. Jeff Weathersby, "Risk Factor in Smelter Emissions Is Mostly Guesswork," *Tacoma News Tribune*, August 7, 1983, H6; and John Gillie, "Asarco Consultant Says Arsenic No Lung Cancer Threat," *Tacoma News Tribune*, October 20, 1983, B3.
122. John Gillie, "Sutherland Calls Asarco 'Good Neighbor' on National TV," *Tacoma News Tribune*, August 10, 1983, A3.

123. Weathersby, "Risk Factor in Smelter Emissions."
124. John P. Sammons, senior vice president, Hill & Knowlton, to Lawrence W. Lindquist, manager, ASARCO Tacoma Plant, July 29, 1983, *Branin v. ASARCO.*
125. Gillie, "Asarco Consultant."
126. Floyd Frost, Lucy Harter, Samuel Milham, Rachel Royce, A. H. Smith, J. Hartley, and P. Enterline, "Lung Cancer among Women Residing Close to an Arsenic Emitting Copper Smelter," *Archives of Environmental Health* 42, no. 2(May–June 1987): 148–152, at 152.
127. Environmental Protection Agency, "Official Transcript of the Proceedings before the Environmental Protection Agency, Volume I, In the Matter of Public Hearing on Proposed Standards for Inorganic Arsenic Emissions: Statement of Dr. Samuel Milham," November 2, 1983, Tacoma, Washington, ASAR ASA219 Frame 0352, 20–31; and ASA219 Frame 0352, 56–66.
128. "Plant Closure: The ASARCO/Tacoma Copper Smelter," in *Environmental Hazards: Communicating Risk as a Social Process*, ed. S. Krimsky and A. Plough, 180–238 (Dover, MA: Auburn House, 1988).
129. "Smelter Putting Arsenic in the Air Is Set to Close," *New York Times*, June 30, 1984, 14; and L. W. Lindquist to Alexandra Smith, EPA Region 10, June 27, 1984, ASDSF 13.4.

CHAPTER 7 SACRIFICED

1. Kevin Rochlin, EPA Site Manager, to Ruston Mayor Hopkins, "Email: Information on ASARCO Project," October 7, 2009.
2. Michael P. Malone, "The Close of the Copper Century," *Montana: The Magazine of Western History* 35 (Spring 1985): 69–72.
3. Environmental Protection Agency Region 8, *Role of Clean Air Act Requirements in Anaconda Copper Company's Closure of Its Montana Smelter and Refinery*, June 24, 1981; and Robert Coughlin to Randy Smith, Region 10 EPA, "Report: ASARCO Closure," January 7, 1985, Washington State Department of Ecology, ASARCO electronic files (hereafter Ecology electronic files).
4. Ernesta B. Barnes, EPA regional administrator, to Charles O'Donohue, business agent Local 25 United Steelworkers of America, March 22, 1985, Environmental Protection Agency Site File, Commencement Bay/Nearshore Tideflats ASARCO Smelter Facility Site File (hereafter ASDSF), 13.1.1.
5. Coughlin to Smith, "Report: ASARCO Closure."
6. Ibid.
7. Dan Voelpel, "Smelter's Fate Sealed by 1975," *News Tribune*, March 29, 1985, Washington State Archives (hereafter WSA), Box 26 Ecology Toxic Cleanup Program 68–86 F: ASARCO press clippings.
8. Barnes to O'Donohue, March 22, 1985.
9. Jeff Weathersby, "Smelter Fallout: Union Asks Probe Continuation," *News Tribune*, March 29, 1985, B2.
10. National Emission Standards for Hazardous Air Pollutants; Standards for Inorganic Arsenic, 51 Fed. Reg. 149 (Aug. 4, 1986), 27956.
11. Arthur Dammkoehler to Donald P. Dubois, EPA Regional Administrator, "Report of Ambient SO_2 Concentrations Exceeding 0.50ppm for a Three-Hour Average (Not to Be Exceeded More Than Once per Year) and Exceeding 0.14ppm for a Twenty-four Hour Average (Not to Be Exceeded More Than Once per Year) Attributable to ASARCO, Inc., Tacoma Smelter, Tacoma, Washington," December 15, 1976, University of Washington Libraries, Special Collections, University Archives, American Lung Association of

Washington Records (hereafter ALA), 5271–001 B3 F56; Armand Labbe to Wilbur E. Hallauer, "ASARCO Tacoma Smelter," January 9, 1979, ALA 5271-1 B3 F52; and Arthur Dammkoehler to Board of Directors, Puget Sound Air Pollution Control Authority, "Memorandum: Status of ASARCO," November 1, 1979, ALA 5271-1 B4 F1.

12. C. E. Dungey to Lynda Brothers, November 4, 1985, Environmental Protection Agency ASARCO Administrative Record on microfilm (hereafter ASAR, with microfilm number and start frame number following), ASA 208 Frame 0290.
13. U.S. EPA, "EPA Sets New National Pollution Standard for Lead," September 28, 1978, http://www.epa.gov/history/topics/caa70/08.html, accessed December 28, 2010; and Pacific Environmental Services, Inc., *Background Report AP-42, Section 12.6, Primary Lead Smelting and Refining* (n.d.), http://www.epa.gov/ttnchie1/ap42/ch12/bgdocs/b12s06.pdf.
14. Katherine G. Aiken, *Idaho's Bunker Hill: The Rise and Fall of a Great Mining Company, 1885–1981* (Norman: University of Oklahoma Press, 2005).
15. "Smelter Can't Live with EPA Rules" *Spokane Chronicle*, February 1, 1985, 4.
16. EPA, "EPA Sets New National Pollution Standard for Lead."
17. Aiken, *Idaho's Bunker Hill*, 195–196.
18. Texas Center for Policy Studies, "Chapter 6: Air Quality," *Texas Environmental Almanac* (1995), http://www.texascenter.org/almanac/Air/AIRCH6P1.HTML, accessed December 30, 2010.
19. Ibid.
20. David Hooper, "City at Bay: The Poisoning of Tacoma," *Pacific Northwest Magazine*, December 1982, 28; and Wallace Turner, "Tacoma Pollution on '10 Worst' List," *New York Times*, November 8, 1981, 35.
21. Dom Reale, CMBA controlled sites project engineer, to Clark Haberman and Phil Miller, "Confidential Memorandum: Recommendation for ASARCO Administrative Order," February 20, 1986, WSA, Ecology, Box 25 Toxics Cleanup Program ASARCO Site '62–86, F: Exempt Correspondence.
22. Becky Kramer, "Ex-Bunker Hill Exec Enters British Politics," *Spokesman Review*, July 14, 2010, http://www.spokesman.com/stories/2010/jul/14/ex-mining-exec-enters-politics/, accessed December 30, 2010.
23. Kenneth A. Konz, acting assistant inspector general for audit, to F. Henry Habicht II, deputy administrator, "Letter Report: Special Review of EPA's Handling of the Bunker Hill Superfund Site, Report Number E6FGGO-13–2005–0400006," January 30, 1990.
24. Environmental Protection Agency, "Bunker Hill Non-Populated Areas First Five-Year Review," October 17, 2000. http://yosemite.epa.gov/R10/CLEANUP.NSF/46453efc0be3985c88256d140050c1ac/d5f88bf8bb21522a88256d0f0070c867/$FILE/5%20yr%20non-pop%201-3.pdf.
25. George W. Andersen, executive vice president, ASARCO, to Charles E. Findley, EPA Region 10, "Ruston Expedited Response Action," May 27, 1988, Environmental Protection Agency Site File, Commencement Bay/Nearshore Tideflats Ruston/North Tacoma Site File (hereafter RNTSF), 4.4.1; Elgin Syferd, "ASARCO Incorporated, Tacoma Smelter Site: Long-Term Communications Plan, Draft," May 20, 1991, *Branin v. ASARCO*, 93-CV-5132, Accession # 021-04-0102, Box 1-9, Location # 3050865, settlement papers obtained at Seattle Federal Records Center (hereafter *Branin v. ASARCO*); Mundy & Associates, "Public Relations Expenditures (Elgin Seyferd)," (n.d., report excerpt) *Branin v. ASARCO*; Elgin Syferd, "Asarco Communications Plan: Asarco Tacoma Remedial Investigation Feasibility Study (Final)," March 2, 1990, *Branin v. ASARCO*; Lori to Dave, "Internal Elgin Syferd Memorandum: ASARCO," October 7, 1991, *Branin v. ASARCO*;

David Marriott to Tom Aldrich, "Memorandum: Media Training," March 10, 1992, *Branin v ASARCO*; and Teresa Stewart and Lori Jarman to Tom Aldrich, "Memorandum: Invitation List for June 11 'Friends' Meeting," May 26, 1992, *Branin v. ASARCO*.

26. The Washington State standard for arsenic in soil for unrestricted use is 20 ppm. Washington State Department of Ecology, Toxics Cleanup Program, "Model Toxic Control Act Statute and Regulation," Publication No. 94-06, November 2007. On the geographic limitations, see Environmental Protection Agency, Region 10, Ruston/North Tacoma Site Preliminary Remedial Action Objectives Decision Memorandum, January 1992, 6, RNTSF 2.6.2.

27. Gordy Holt, "How Much Is Too Much Arsenic? Fear Poisons Lives of Many Residents," *Seattle Post-Intelligencer*, September 11, 2000, A1.

28. Gregory L. Glass, *Credible Evidence Report: The ASARCO Tacoma Smelter and Regional Soil Contamination in Puget Sound: Final Report*, September 2003, Prepared for Tacoma-Pierce County Department of Health and Washington State Department of Ecology.

29. See, for example, Brian Everstine, "Make Asarco Pay for Environmental Cleanup Says Cantwell," *Tacoma News Tribune*, August 6, 2008.

30. U.S. EPA, "American Smelting and Refining Company (ASARCO) Bankruptcy Settlement," http://www.epa.gov/compliance/resources/cases/cleanup/cercla/asarco/index.html, accessed September 5, 2010.

31. Washington State Department of Ecology, "News Release: ASARCO Settlement Money Funds New Soil Clean-ups," June 30, 2010, http://www.ecy.wa.gov/news/2010news/2010-148.html, accessed July 29, 2012.

32. Committee on Superfund Sites Assessment and Remediation in the Coeur d'Alene River Basin, *Superfund and Mining Megasites: Lessons from the Coeur D'Alene River Basin* (Washington DC: National Academies Press, 2005); Environmental Protection Agency, "Record of Decision, Bunker Hill Mining and Metallurgical Complex OU3," September 2002; and Environmental Protection Agency, "Press Release: Asarco Settlement and Bunker Hill Superfund Cleanup: Frequently Asked Questions," March 2010, http://yosemite.epa.gov/R10/CLEANUP.NSF/sites/bh, accessed July 7, 2010.

33. See, for example, Robert McClure, "Town Split over Direction of Cleanup," *Seattle Post-Intelligencer*, October 21, 2001, http://www.seattlepi.com/local/41086_silver02.shtml, accessed July 14, 2010; and Paul Koberstein, "Idaho's Sore Thumb," *Cascadia Times*, Spring 2002, http://www.times.org/archives/2002/thumb1.htm, accessed July 14, 2010.

34. Environmental Protection Agency, "Superfund Program Implements the Recovery Act," http://www.epa.gov/superfund/eparecovery/bunker_hill.html, accessed January 16, 2011.

35. S. M. Moodie, E. K. Tsui, and E. K. Silbergeld, "Community- and Family-level Factors Influence Care-giver Choice to Screen Blood Lead Levels of Children in a Mining Community," *Environmental Research* 110 (July 2010): 484–496; Idaho Department of Health and Welfare, Idaho Department of Environmental Quality, Panhandle Health District, U.S. Environmental Protection Agency, "Presentation: 2011 Bunker Hill Superfund Site, Couer d'Alene Basin Blood Lead Levels," February 15, 2012.

36. Moodie, Tsui, and Silbergeld, ""Community- and Family-level Factors."

37. EPA, "American Smelting and Refining Company (ASARCO) Bankruptcy Settlement, EPA Funded Sites and Communities," http://www.epa.gov/compliance/resources/cases/cleanup/cercla/asarco/community.html, accessed July 29, 2012.

38. EPA, Region 10, *First Five-Year Review Report for Ruston/North Tacoma Superfund Site: Ruston and Tacoma, Washington*, March 2000, RNTSF 6.9; and Idaho Department of Environmental Quality, "Mine Waste Management in Idaho: Bunker Hill Superfund Site, Soil

Repositories in the Basin," http://www.deq.idaho.gov/regional-offices-issues/coeur-dalene/bunker-hill-superfund-site/soil-repositories.aspx, accessed July 29, 2012.

39. L. Stokes, R. Letz, F. Gerr, M. Kolczak, F. E. McNeill, D. R. Chettle, and W. E. Kaye, "Neurotoxicity in Young Adults 20 Years after Childhood Exposure to Lead: The Bunker Hill Experience," *Occupational Environmental Medicine* 55 (1998): 507–516; Z. Berkowitz, P. Price-Green, F. J. Bove, and W. E. Kaye, "Lead Exposure and Birth Outcomes in Five Communities in Shoshone County, Idaho," *International Journal of Hygiene and Environmental Health* 209 (2006): 123–132; and Nicholas K. Geranios, "Victims of North Idaho Lead Poisoning Still Suffer Physical, Emotional Ills," *Missoulian*, September 5, 2010, http://missoulian.com/news/state-and-regional/article_49524244-b911-11df-accd-001cc4c002e0.html, accessed March 26, 2012.

40. Sean Sheldrake, project manager, EPA Region 10, to File, "Memorandum: August 12 Mining Company Presentation to State Policy Group—Coeur d'Alene Basin Human Health Risk Assessment and Bunker Hill ATSDR Medical Monitoring Program," August 13, 1999, EPA Superfund Records, Bunker Hill Site File (hereafter BHSF), 16.5.8; Richard H. Schultz, administrator, Division of Health, to Karl B. Kurtz, director, Department of Health, "Memorandum: Recommendation on Bunker Hill Monitoring Proposal from Governor's Basin Task Force," November 18, 1999, BHSF 16.5.8; and Karl B. Kurtz, director IDHW, to Governor Kempthorne, "State's Role in Implementing the Proposed Medical Monitoring in the Coeur d'Alene River Basin," December 8, 1999, BHSF 16.5.8; and Zaz Hollander, "Idaho Rejects Bunker Hill Health Tests," *Idaho Spokesman Review*, November 10, 2000, http://www.spokesmanreview.com/news-story.asp?date=111000&ID=s877964, accessed December 28, 2010.

41. ATSDR, "El Paso Multiple Sclerosis Cluster Investigation, El Paso, El Paso County, Texas Final Report," August 2001.

42. "ASARCO Hayden Plant, EPA Superfund Alternative Site," December, 2009, http://www.azdeq.gov/environ/waste/sps/download/state/asarco.pdf, accessed January 7, 2011.

43. John M. Broder, "EPA Issues Limits on Mercury Emissions" *New York Times Green Blog*, December 21, 2011, http://green.blogs.nytimes.com/2011/12/21/e-p-a-announces-mercury-limits/, accessed June 15, 2013.

44. Charles W. Schmidt, "Lead in Air: Adjusting to a New Standard," *Environmental Health Perspectives* 118 (February 2010): a76–a79; EPA National Emissions Inventory (2008), http://www.epa.gov/ttn/chief/net/2008inventory.html, accessed January 26, 2013; and "Approval and Promulgation of Implementation Plans; State of Missouri," 75 Fed. Reg. 166 (August 27, 2010), 52701.

45. Schmidt, "Lead in Air."

46. "ASARCO Hayden Plant"; and CH2MHill, "Baseline Human Health Risk Assessment for the ASARCO LLC Hayden Plant Site, Hayden, Gila County, Arizona," August 2008. For example, estimated excess lifetime cancer risk from air in Hayden for arsenic, cadmium, and chromium is 1 in 10,000, with 80 percent of the risk attributable to arsenic.

47. Judy Fahys, "Big Mines Linked to Increase in Toxic Releases in 2010," *Salt Lake Tribune*, January 6, 2012.

48. Judy Fahys, "Utah Pollution Fighters Sue Kennecott," *Salt Lake Tribune*, December 19, 2011.

49. EPA, "Our Nation's Air: Status and Trends through 2010," February 2012, EPA-454/R-12-001, http://www.epa.gov/airtrends/2011/report/coverandtoc.pdf, accessed April 11, 2012.

50. EPA, "Summary of Results for the 2005 National Scale Assessment" (2011), http://www.epa.gov/ttn/atw/nata2005/, accessed April 11, 2012.

51. Jim Morris, "Poisoned Places: Toxic Air, Neglected Communities," *Center for Public Integrity*, http://www.iwatchnews.org/2011/11/07/7267/many-americans-left-behind-quest-cleaner-air; and Howard Berkes, "EPA Takes Action against Toxic Arizona Copper Plant," *National Public Radio*, http://www.npr.org/blogs/thetwo-way/2011/11/17/142439081/epa-takes-action-against-toxic-arizona-copper-plant, accessed April 11, 2012.
52. EPA, "Fact Sheet: Mercury and Air Toxics Standards" (n.d.), http://www.epa.gov/mats/pdfs/20111221MATSimpactsfs.pdf, accessed April 4, 2012; and EPA, Office of Air and Radiation, "The Benefits and Costs of the Clean Air Act from 1990–2020, Summary Report," March 2011.
53. International Copper Study Group, *The World Copper Factbook* (2010), available at http://www.scribd.com/doc/52153185/2010-World-Copper-Factbook.
54. Blacksmith Institute, *The World's Worst Polluted Places: The Top Ten of the Dirty Thirty* (New York: Blacksmith Institute, September 2007); and Andrew E. Kramer, "A Blighted Russian City Pans for Cash," *New York Times*, July 11, 2007, http://www.nytimes.com/2007/07/11/world/europe/11iht-letter.4.6616383.html, accessed January 31, 2011.
55. M. F. Khokhar, U. Platt, and T. Wagner, "Temporal Trends of Anthropogenic S02 Emitted by Non-ferrous Metal Smelters in Peru and Russia Estimated from Satellite Observations," *Atmospheric Chemistry & Physics Discussions* 8 (September 18, 2008): 17393–17422.
56. Blacksmith Institute, *World's Worst Polluted Places*.
57. Centers for Disease Control and Prevention, National Center for Environmental Health/Agency for Toxic Substances, and Disease Registry Division of Emergency and Environmental Health Services, *Development of an Integrated Intervention Plan to Reduce Exposure to Lead and Other Contaminants in the Mining Center of La Oroya, Peru* (May 2005).
58. See, for example, "Lead Poisoning Hits 84 Chinese Kids Near Smelters," *Reuters*, July 26, 2010, http://www.reuters.com/article/2010/07/26/us-china-lead-poisoning-idUSTRE66P1OO20100726, accessed September 5, 2010; "'Hundreds Ill' near China Smelter," *BBC*, August 20, 2009, http://news.bbc.co.uk/2/hi/8211081.stm, accessed September 5, 2010; Sharon LaFraniere, "Lead Poisoning in China: The Hidden Scourge," *New York Times*, June 15, 2011, http://www.nytimes.com/2011/06/15/world/asia/15lead.html; Lucy Hornby, "Lead Poisoning Haunts Chinese Smelter Communities," *Reuters*, August 24, 2009, http://www.reuters.com/article/2009/08/24/us-china-pollution-idUSTRE57N16B20090824, accessed January 11, 2013; Jonathan Watts, "1300 Chinese Children Near Smelter Suffer Lead Poisoning," *Guardian*, August 20, 2009, http://www.guardian.co.uk/world/2009/aug/20/china-smelter-children-lead-poisoning, accessed January 11, 2013; "Smelter to Fully Close after Lead Poisoning Sickens 851 Kids," *China Daily*, August 20, 2009, http://www.chinadaily.com.cn/business/2009-08/20/content_8594661.htm, accessed January 11, 2013; and Human Rights Watch, *My Children Have Been Poisoned: A Public Health Crisis in Four Chinese Provinces*, June 15, 2011, http://www.hrw.org/reports/2011/06/15/my-children-have-been-poisoned-0.

CONCLUSION

1. Sharon LaFraniere, "Lead Poisoning in China: The Hidden Scourge," *New York Times*, June 15, 2011, A1.
2. For more on industry influence on regulation and regulators, see, for example, Robert Galbraith, *Fracking and the Revolving Door in Pennsylvania*, Public Accountability Initiative, February 2013, http://public-accountability.org/wp-content/uploads/

Fracking-and-the-Revolving-Door-in-Pennsylvania.pdf, accessed June 10, 2013; Lee Fang, "New ALEC Documents Show Regulatory Capture in Action," *The Nation*, December 19, 2012, http://www.thenation.com/blog/171852/new-alec-documents-show-regulatory-capture-action#axzz2W8NkYid9, accessed June 10, 2013; Ian Urbina, "Pressure Limits Efforts to Police Drilling for Gas," *New York Times*, March 3, 2011, http://www.nytimes.com/2011/03/04/us/04gas.html?pagewanted=all&_r=0, accessed June 10, 2013; John M. Broder, "Panel Urges Tougher Offshore Regulation," *New York Times Green Blog*, September 8, 2010, http://green.blogs.nytimes.com/2010/09/08/panel-urges-tougher-offshore-regulation/, accessed June 10, 2013; and Bruce Barcott, "Changing All the Rules, " *New York Times*, April 4, 2004, http://www.nytimes.com/2004/04/04/magazine/04BUSH.html?pagewanted=all, accessed June 10, 2013. On the role of the OMB and the White House in environmental regulation, see Lisa Heinzerling, *Who Will Run the EPA? Yale Journal on Regulation*, March 27, 2013; Rena Steinzor, Michael Patoka, and James Goodwin, *Behind Closed Doors at the White House: How Politics Trumps Protection of Public Health, Worker Safety, and the Environment*, Center for Progressive Reform White Paper #1111, November 2011, http://www.progressivereform.org/articles/OIRA_Meetings_1111.pdf, accessed June 10, 2013; and Andrew Revkin and Matthew Wald, "Material Shows Weakening of Climate Reports" *New York Times*, March 20, 2007, http://www.nytimes.com/2007/03/20/washington/20climate.html?_r=0, accessed June 10, 2013.

3. John M. Broder, "Bashing E.P.A. Is New Theme in G.O.P. Race," *New York Times*, August 17, 2011, http://www.nytimes.com/2011/08/18/us/politics/18epa.html, accessed March 25, 2012.

4. Allen M. Brandt, *The Cigarette Century: The Rise, Fall, and Deadly Persistence of the Product That Defined America* (New York: Basic Books, 2007).

5. Gerald Markowitz and David Rosner, *Deceit and Denial: The Deadly Politics of Industrial Pollution* (Berkeley: University of California Press, 2002).

INDEX

Page numbers followed by an f indicate a figure.

acceptable risks, 148
acid rain, 3; smelter's contribution to, 142; and sulfur dioxide emissions, 15
adaptive abilities, in lead-exposed children, 205n174
Agency for Toxic Substances and Disease Registry (ATSDR), 120, 163
agriculture, use of arsenic in, 115–116
Aiken, Katherine, 79
air pollutants: EPA standards for, 149; federal standards for, 62; standards for hazardous, 131. *See also* arsenic; lead; sulfur dioxide
air pollution, 168, 173; cleaning up hazardous, 167; conditions associated with, 36–37; control of research on, 33; deaths resulting from, 35, 36, 42, 48; estimated reduction in, 167; health effects of, 29, 35, 48, 55; public concerns about, 48; research on health effects of, 168; in Tacoma, WA, 27–28, 34. *See also* arsenic; lead; sulfur dioxide
air pollution control, 47; ASARCO's delays in, 156; local, 45; public demands for, 82; regulatory authority for, 35; in U.S., 51
air quality, in Tacoma, WA, 36
air sampling data, for El Paso area, 63
air toxics: cleaning up, 167; health benefits of reducing, 168; as public health issue, 167. *See also* toxic metal emissions
Albert, Roy, 139
Allen, Robert H., 80, 87, 89–90, 105
Alsos, Judy, 41, 43, 46, 47–48
Amalgamated Copper Co., 3
Amarillo, TX, ASARCO refinery in, 156
ambient air, federal standards for lead in, 104, 105
ambient lead levels, seasonal fluctuation in, 96–97, 102. *See also* blood lead levels
Amdur, Mary, 41–42
American Lung Association, Washington chapter of, 134

American Public Health Association, 89
American Smelting and Refining Co. (ASARCO), 1, 3, 58; and airborne lead, 65; alleged dumping by, 140; and arsenical dermatoses, 210n37; and arsenic control, 27; and arsenic exposure, 112; and arsenic standard, 156–157; and arsenic testing in 1972, 124–125; bad publicity feared by, 124; bankruptcy of, 160; and Bunker Hill Mining Co., 71; and cleanup costs, 159, 160; contaminated water discharged by, 141; crisis management of, 70; damage payments of, 50, 61; effective strategies of, 26; and El Paso incident, 55, 57, 66; "environmental progress" of, 124; environmental record of, 13; EPA negotiations with, 144–145; internal report on El Paso, 73; and lead emission crisis, 64; liability fear of, 145; liability for toxic trespass of, 146; media support of, 148; meeting with Bunker Hill/Gulf, 89; nonferrous smelters of, 13; political support sought by, 48–49; press release by, 135; public relations, 38, 123–124, 151, 159, 177n18; resistance of, 54; safety claims made by, 29; sealed internal documents of, 165; soil theory of, 64–66; on sulfur dioxide, 18, 19, 179n32; sulfur dioxide control attempts of, 180n35; and sulfur dioxide emissions, 124; and sulfur dioxide standard, 157; sulfur emission capture of, 37; Tacoma closure anticipated by, 156; and Tacoma's community organizing, 44; Tacoma smelter closure announced by, 152; Tacoma smelter purchased by, 11; taller smokestacks promoted by, 17; threshold argument used by, 137, 150; urinary arsenic testing conducted by, 134; on workplace arsenic standard, 132
American Smelting and Refining Co. (ASARCO), lawsuits against: Mike and

227

(ASARCO), lawsuits against (*continued*)
  Marie Bradley's, 145–146; El Paso's, 59, 62, 65; David Reed's, 145; Tacoma residents', 164; unsuccessful, 145–146; Wingards', 144
American Smelting and Refining Co. (ASARCO) research: arsenic study, 117; cancer risk study, 136–137; Dept. of Agricultural Research of, 18; invisible injury theory funded by, 180n40; research funded by, 19, 29; science sponsored by, 6, 7, 13–14; on urinary arsenic concentrations, 120; Utah Research Dept. of, 25
American Society of Mechanical Engineers, 36
Anaconda Co., 22; corporate structure of, 16; damage payments of, 16, 17; taller smokestacks promoted by, 17
Anaconda smelter, in Montana, 3, 16; closure of, 147; and risk of lung cancer, 118; workers at, 106
Anapra, Ciudad Juarez, Mexico, 70
Anapra, NM, 58
Andersen, Carl and Frederica, 22
aniline dye workers, cancer in, 116, 208n17
animals, arsenic poisoning of, 21, 210n37.
  See also livestock; pets
APE (Americans Protecting the Environment), 134
"aroma of Tacoma," 32, 39
arsenic: ambient air concentrations of, 126, 132–133; anthropogenic, 11, 23; ASARCO study of, 117; in bees, 143; carcinogenicity of, 5, 113, 114, 117, 118, 119, 125, 131, 139; in dairy cows, 130, 213–214n5; detection of, 118; dust, 121, 144, 210n37; federal standard for, 139; in Hayden, AZ, air, 166; lung cancer associated with, 116; market for, 27; medicinal use of, 114–115, 116; NESHAP for, 147; OSHA's standard for, 132; in pesticides, 24, 119; pollution, 125; profitability of, 23, 26; properties of, 23; recovery of, 20; regulation of, 131, 139; research on, 118–119; role in carcinoma contested, 117; science on environmental, 145–146; sources of, 23; standards for, 7; toxicity of, 22–23, 23–24; urinary, 210n34 (*see also* urinalysis screening)
arsenical dermatoses: in children, 210n37; urinary arsenic concentrations in, 123
arsenic capture, decisions regarding, 183n85
arsenic control: with electrostatic precipitators, 27–28; in Tacoma, WA, 25
arsenic emissions: compared with lead emissions, 138; follow-up studies of children exposed to, 163; remediating contamination with, 160; from Tacoma smelter, 21
arsenic exposure: ASARCO's categorization of, 211–212n1; and ASARCO's PR strategy, 151; cancers associated with, 22; of children, 2, 133, 138, 157; community, 121; dermatologic manifestations of, 210n37; fugitive emissions, 29; health effects of, 46; human health consequences of, 112; inappropriate categorization of exposure status in ASARCO study, 123; lack of public health guidelines for, 119, 120; and livestock, 28, 130, 213–214n5; and lung cancer, 132; in medicine, 115; occupational, 115, 116; of Ruston children, 134; in Ruston, WA, 13, 130; for smelter workers, 118; in Tacoma, WA, 111, 112, 126; and threshold concept, 137; uncertainty about, 140; on Vashon Island, 130
arsenic industry, 120
arsenicism, 114, 115
arsenico-dermia, 115
arsenic poisoning, 24; "benign," 120; low-level, 119; mass, 24; signs of, 114
arsenic regulations, ASARCO's compliance with, 137
arsenic residue, on food, 117
"arsenic showers," 21, 29, 45–46
arsenic soil contamination, in Puget Sound Region, 161f
arsenic trioxide (As2O3), 23, 117
ASARCO. *See* American Smelting and Refining Co.
Asiatic pills, 114, 115
asthmatics, and sulfur dioxide exposure, 41
attention deficit hyperactivity disorder, and lead exposure, 164

baghouses, 19; at Bunker Hill smelter, 82, 83–84, 95; costs associated with, 20; fire in, 82, 164
Baker, Gene M., 81, 83, 84, 89, 92, 94, 99, 104
Bangladesh, arsenic poisoning in, 23
Barber, Charles F., 43, 48–49, 73, 146, 156
Barltrop, Donald, 67, 92
Bax, James A., 85, 90, 91
bees: arsenic toxic to, 22; heavy metal concentrations in, 143; and Tacoma smelter, 112
Bergner, Lawrence, 213–214n5
"best available technology" standard, to control emissions, 148
Biden, Sen. Joe, 68
Birmingham, D. J., 210n37
blister copper, 15
blood lead levels (BLLs): in Herculaneum, MO, 166; in Kellog, ID, 4; long-term health problems associated with, 93; and NCV, 98, 99; seasonal fluctuation in, 96–97, 102; in Shoshone Project, 93; in Silver Valley, ID, 162; in Smeltertown children, 177n18; Tacoma smelter's contribution to elevated, 122; threshold level for, 60; in U.S., 100; on Vashon Island, 214n5
blood lead sampling, ongoing, 88
bone lead, 84
Boyd, William, 90
Bradley, Mike and Marie, 145–146
brain dysfunction, in lead-exposed children, 205n174
Brainerd, Wendell, 96
breathing problems, in Tacoma children, 41
broken back syndrome, in herring, 141

INDEX 229

Bromenshenk, Jerry, 143
Bucove, Bernard, 47
Bunker Hill disaster, public discourse controlled in, 74–75
Bunker Hill Mining Co., 71, 73; annual testing controlled by, 100; collaboration with State Health Dept., 102–103; community response to, 97–98; crisis response of, 88–89; damage from lead poisoning denied by, 98; and Doctor's Clinic, 87, 196–197n12; government support of, 105; grant offered to state by, 101; "greening program" of, 96; and ILZRO, 89, 90; investigation of, 107; medical care funded by, 90; ongoing claims made by, 95; planned independent study by, 89; public relations strategies of, 95–96, 99; reaction to blood lead levels of, 100; relationship with Doctor's Clinic, 75; relationship with local doctors, 109; residential homes demolished by, 94; science sponsored by, 6, 7; sealed internal documents of, 165; Shoshone Project funded by, 6, 176n14; successful lawsuit against, 164
Bunker Hill smelter, 4, 70, 73–74, 75f; and air quality standards for lead, 157; baghouse fire at, 82, 164; closure of, 105, 110, 157; compared with Tacoma's smelter, 81–82; compliance schedule for, 84; damage complaints against, 79; damage to key pollution control device at, 82, 83; fifty-year anniversary of, 79f; pollution control at, 77; resistance to pollution control of, 82
Bunker Hill smelter emissions, 77, 87; Areas I and II, 87; and blood lead levels, 85; children's exposure to, 87; damage to vegetation from, 81; fugitive, 95; heavy metals, 81; and Nixon administration, 80; in Shoshone Project, 93
Bunker Limited Partnership (BLP), 159
Bunker Smelter/Gulf, and ILZRO, 201n96
Bureau of Mines, U.S., 14; and ASARCO's research, 18–19; and mining and smelting industry, 16; on toxic metal recovery, 20
Bush (George H. W.) administration, 167
Busselle, B. B., 34

cadmium: children exposed to, 108; in smelter emissions, 122
cadmium exposure, 93
California, air lead standard of, 104
California Supreme Court, 27
cancer: and arsenic exposure, 5, 22; respiratory mortality from, 116, 117; scrotal, 116; skin, 114–116. *See also* lung cancer
cancer mortality, and arsenic exposure, 132
cancer risk: adult *vs.* childhood, 137–138; from arsenic, 131–132, 148; as defined by community, 152; for Tacoma residents, 148, 151
capture technology, for sulfur dioxide, 18, 37
Carcinogen Assessment Group (CAG), EPA's, 139

carcinogenesis, arsenic, 116
carcinogens, 113, 114, 119, 121
cardiovascular disease, and bone lead, 84
Carnow, Bertram, 62, 64, 77, 87
Carson, Rachel, 24, 51
Casner, Nicolas, 78
catalytic converters, 60
CDC. *See* Centers for Disease Control and Prevention
Center for Public Integrity, 167
Centers for Disease Control and Prevention (CDC): air sampling data of, 64; and blood lead levels in Idaho, 101–102; Bunker Hill/Gulf's opposition to, 89–90; and community health effects from smelters, 71; El Paso investigation of, 59, 70; focus on vaccine preventable illness of, 62; and investigation of Bunker Hill emissions, 87; 1974 study of, 98; and Shoshone Project, 100
Chamber of Commerce, Tacoma, 37
chelation treatment, 68; hospitalization for, 85; outpatient *vs.* hospital, 108; risk associated with, 88
children: arsenic exposure of, 121, 133, 134–136, 216n31; blame for contamination placed on, 136; effects of lead on nervous systems of, 88; in El Paso elementary school, 158; Fern Hill Elementary School, 121; health effects of arsenic on, 163; lawsuit of lead poisoned, 83; lead exposure of, 59, 60, 81; lead poisoning of, 4–5, 63–64, 66, 74; near Bunker Hill smelter, 84; at Silver King Elementary School, 76–77, 94–95, 103, 108; subclinical effects of lead on, 67, 70; urinary arsenic concentrations in, 121
children, testing: for cognitive deficits, 211n49; on Vashon Island, 213–214n5. *See also* blood lead levels; urinalysis
China: environmental concerns in, 9; lead poisoning of children in, 169, 171
Chisolm, J. Julian, 67, 76, 92
citizen complaints, in Tacoma, 32
citizen groups, 134. *See also* community organizing
citizens, need for engaged and educated, 171
Citizens' Committee on Air Pollution, Tacoma, WA, 53; founding of, 43; petition presented by, 46; and Washington State policies, 46
City of Destiny, 31. *See also* Tacoma
Ciudad Juarez, Mexico, 58, 70
clean air: achieving, 167; public enthusiasm for, 113
Clean Air Act, federal (1970), 54, 55, 60, 113; limited progress with, 167; 1990 revisions to, 167, 168; section 112 of, 149; standards for hazardous air pollutants in, 149
Clean Air Act, federal (1977), 139; and sulfur dioxide emission regulations, 142
Clean Air Act, Washington State (1968), 53
Clean Air for Washington, 134

INDEX

cleanup: in El Paso, 162; for hazardous air pollution, 167; limits of, 162–163; in Ruston, 160; Silver Valley, 160, 164
climate, and air pollution, 81, 198n45
climate change, 168; effective regulatory action on, 173
Coast Guard, U.S., 23
Coeur d'Alene mining district, 77, 78f
Coeur d'Alene Press, 96
cognitive deficits in children, testing denied for, 211n49
Cole, Jerome F., 58, 69, 70, 89, 90, 91, 92, 104, 194n81
communities: and air toxics, 167–168; devastated by toxic waste, 165; fenceline, 114; in Kellogg, ID, 97; reaction to trials of, 110; role of, 110
communities, smelting, 158; political activity of, 5–6; public health of, 5; resistance of, 6; sharp divisions in, 7
Community Air Pollution Demonstration Project in Tacoma, proposal for, 43–45
community environmental health, 172
community health: and ambient arsenic, 141; ambivalence toward, 114; corporate impact on, 142; *vs.* economic interests, 138; economics of, 8, 138; and industrial pollution, 71, 171; and industry power, 174; near nonferrous smelters, 70; and Tacoma smelter, 140
community organizing, 36; for pollution control, 29; and reduction in emissions, 133–134; for tighter controls on smelter emissions, 152
community organizing, in Tacoma, WA, 43; and city officials, 53; opposition to, 32; against smelter pollution, 36
confidentiality, in Shoshone project, 94
conflict of interest, in Shoshone project, 94
consent decree, ASARCO–El Paso–State of Texas, 65, 192n45
Conservation and Natural Resources Subcommittee, House, 80
consumer products, arsenic in, 147
Cooper, Clark, 104
copper industry, 26
copper matte, 15
copper ore: high-arsenic, 22; smelted at Tacoma, 130
copper smelting, sulfur dioxide in, 15. *See also* smelting
Cordwell, Robert, 88
corporate culture, at Bunker Hill smelter, 79–80
cost-benefit analysis: decision making, 149; in risk assessment, 149–150
Costle, Douglas, 105
costs: of baghouses, 20; of pollution control, 50, 168
Cottrell electrostatic precipitators, 20; and arsenic control, 27–28. *See also* electrostatic precipitators
court records, access to, 5

courts: and chronic pollution, 131; as last-ditch solution, 165; lead poisoning cases in, 109
crisis management, 70
crop damage: ASARCO's research on, 19; from sulfur dioxide emissions, 15, 16; in Tacoma, 25. *See also* soils, vegetation
Cutchins, Bob, 86

Dahlberg, Dr., 84
damage payments: from ASARCO, 50, 61; *vs.* pollution controls, 3
Dammkoehler, Arthur, 134
Davis, Devra, 42
death rate, and urban air pollution, 36–37
Dekan, George, 74, 83
Delaney Clause, 117
Department of Environmental and Community Services (DECS), 83, 85
Depression, pollution control during, 33
dermatoses: arsenical, 140, 210n37; in children, 210n37
DiGiacomo, Ronald, 138
disasters: air pollution, 35, 36. *See also* Bunker Hill disaster; El Paso
Dixon, Fritz, 102
Doctor's Clinic, in Kellogg, ID, 75, 87, 196–197n12; Bunker Mining Company's medical plan with, 87, 135, 196–197n12
Doe Run Co., 166
Donora, PA, air pollution deaths at, 35, 36, 42, 48
Donovan's Solution, 114
doubt, in industry's argument, 172. *See also* manufactured doubt
Drinker, Philip, 41–42
Ducktown smelter, 18
Dungey, Curtis, 135, 136
DuPont Co., and El Paso research, 67
Durnin, Joel, 44, 45, 47
dust: arsenic, 121, 144, 210n37; from Tacoma smelter, 144

East Helena smelter, 50, 157
economics, of community health, 8, 138
economy, impact of arsenic exposure on, 112
electric utilities, compliance of, 165
electrostatic precipitators, 20; and arsenic control, 27–28; claims for, 27–28; at Tacoma smelter, 25–26, 27
El Paso, TX, 2; ASARCO consent decree with, 65, 192n45; heavy-metal emissions in, 56; Kern Place neighborhood, 63; lead emissions in, 59; lead exposure in, 61; lead poisoning in, 54, 57, 111; location of, 55, 57f; residential soil contamination in, 162
El Paso Board of Health, 66
El Paso County Medical Society, Lead Surveillance Committee of, 64
El Paso crisis, compared with Bunker Hill's, 86
El Paso elementary school, arsenic concentrations in, 158

INDEX

El Paso smelter, 4; closure of, 96, 157–158; heavy-metal emissions of, 588; impact on Mexican children, 70; toxic metal emissions of, 157–158

El Paso study (McNeil study), 66, 67–68, 69, 89, 98, 176n14; funding of, 6, 176n14; ILZRO-backed, 91

emergency, smelter pollution as, 53

emission control, state's role in, 101

emissions, smelter, 2; fugitive, 28–29, 126; industrial compared with residential, 39; makeup and magnitude of, 14; reductions in, 133–134

encephalopathy, lead, 77

Enterline, Philip, 68–69, 92, 136

environment, citizen concern about, 32

Environmental Defense Fund, 139

environmental health: children's, 163; political discourse on, 173

environmental impact statement, on Tacoma smelter's continuing operation, 145

environmentalism, 52

environmentalists: and arsenic exposure, 112; divisiveness between workers and, 156

environmental law, 143–144

environmental pollution, and community response, 110. *See also* community organizing

Environmental Protection Agency (EPA), U.S.: and air toxics regulation, 167; and ambient air arsenic, 132–133; antigovernment sentiment aimed at, 162; and antiregulatory backlash, 165; and arsenic exposure, 112, 150; and arsenic regulation, 113; and arsenic's carcinogenicity, 131, 139; and ASARCO cleanup, 159; behind-the-scenes negotiations with ASARCO, 144–145; Bunker Hill agreement with, 82; and Bunker Hill smelter, 80; CAG of, 139; and closure of Tacoma smelter, 156; established, 21, 55; inaction of, 156; industrial pressure on, 104, 105; lead standard of, 105; on McNeil and Landrigan studies, 69–70; mismanagement of, 147–148; national smelter study of, 138; during Reagan administration, 140; and regulation of gasoline, 60; and smelter's health risks, 141; as source of controversy, 173–174; and sulfur dioxide standards, 56, 142

environmental regulations, 4

environmental research, 113. *See also* research

environmental science, 126

EPA. *See* Environmental Protection Agency

Epidemiologic Intelligence Service, 61

epidemiologic studies: of arsenic in air, 141; community health in, 171

Ethyl Corp., 92; El Paso research funded by, 66

Europe, arsenical cancer recognized in, 117

expert witnesses, for ASARCO, 22, 125

Faning, E. Lewis, 116, 117

farmers: disputes with smelting companies of, 17; pollution complaints of, 78–79

federal government: air pollution research supported by, 35; and air pollution standards, 62; and Bunker Hill disaster, 74–75

federal government, role of, 32, 51; in air pollution, 55; and allegations of overregulation, 157. *See also* Environmental Protection Agency

Fern Hill Elementary School, 121

Fetterolf, E. R., 44, 45

"FODOR," 39

Foege, William, 101–102

Fogg, Elizabeth Metcalf, 34

forest damage: from Anaconda smelter, 17; from sulfur dioxide, 16

Fowler's Solution, 114, 115

Friends of the Earth, 134

fugitive emissions: from Bunker Hill smelter, 95; defined, 28–29; estimates for, 126

Fukushima plant, radiation from, 173

Gallagher, Edward, 101, 102, 108

garden soils, smelter contamination of, 143. *See also* soils

Garfield, pilot plant at, 180n35

Garfield copper smelter, 20

Gartside, P. S., 98, 99

gasoline, lead phased out of, 59, 100, 104

GASP (Group against Smog Pollution), 134

Gehlbach, Stephen, 63

Germany, pesticide use in, 24

Glass, Gregory L., 11

Glover smelter, 157

Gorsuch, Anne, 140, 147, 149

government: indifference of, 169; and pollution control, 14. *See also* federal government; state government

greenhouse gas emissions, 173; efforts to reduce, 8. *See also* sulfur dioxide

Greenland, historical evidence of smelting in, 15–16

Greenpeace, 52

Guggenheim family, 3, 11

Gulf Resources and Chemical Corp., 77, 79–80, 87, 105, 109; bankruptcy declared by, 158; and closure of Bunker Hill operations, 110; crisis response of, 88–89; and ILZRO, 89, 90; sealed internal documents of, 165

Haar, Gary Ter, 92

hair lead levels, 81

Halley, James, 109

Hamilton, Alice, 120

Hammond, Paul, 61, 92

Hardy, Harriet, 120

Harkins, William D., 17

Harvard University, ASARCO funded research at, 41

Hayden, AZ, ASARCO smelter in, 156, 165, 166

health, human, 173; in American life, 8; *vs.*

health, human (*continued*)
  economic gains, 8; and sulfur dioxide emissions, 15. *See also* community health
health issue: air pollution as, 36; smelter pollution as, 36
heavy-metal emissions: from Bunker Hill smelter, 81; in El Paso, 56, 58, 70; from Tacoma smelter, 122
heavy-metal exposure, 2; long-term effects of, 107–108; and neurological diseases, 164; science of, 5; in Tacoma, 111
heavy metal intoxication epidemic, 95
heavy metals: arsenic, 23; from El Paso's smelter emissions, 58, 70; fears about, 123; in Hayden, AZ, air, 166; in pubic spaces, 158; in smelter communities, 158; in smelter emissions, 61–62. *See also* lead; mercury
Hecla (mining companies), 164
Herculaneum, MO, lead standard in, 166
Herculaneum smelter, 157
herring eggs, off Vashon Island, 141
Higgins, Ian, 151
Hill, A. Bradford, 116, 117
Hill and Faning study, 117
Hill & Knowlton (public relations firm), 68, 89, 151
Hooker Chemical Co., 38
Hopkins, Bruce, 155
hospitalization: for lead poisoning, 82. *See also* chelation therapy
Hueper, Wilhelm, 24, 115
Human Rights Watch, 169
Hutchinson, Sir Jonathan, 114
hypertension, and lead exposure, 164

Idaho Mining Association, 97
Idaho State: air monitoring by, 81; Air Pollution Commission of, 82; and Bunker Hill disaster, 74–75; and EPA standards, 105; mining in, 80
Idaho State Health Dept., 87; annual testing controlled by, 100; and Bunker Hill's pollution, 95; and CDC, 101–102; CDC banned by, 90; collaboration with Bunker Hill Mining Co., 102–103; damage from lead poisoning denied by, 98; editorial protest aimed at, 97; and follow-up screening, 164; resistance of, 99, 203–204n146. *See also* Shoshone Project
Idaho State Journal, 97
ILZRO. *See* International Lead Zinc Research Organization
Industrial Hygiene Foundation, 120
industrial hygiene research, 41–42
industrial pollution: changing attitudes toward, 50; and community health, 171; ecological consequences of, 173; public intolerance for, 38; record of, 5
Industrial Toxicology (Hamilton and Hardy), 120
industry: and air pollution control, 51; cooperation with universities of, 38; indifference of, 169, 174; and pollution control, 14; prioritization of, 174; use of arsenic in, 115. *See also* mining and smelting industry; smelting industry
infant mortality: in Pierce County, 47; and sulfur dioxide air pollution, 42; and urban air pollution, 36–37
infants, lead exposure of, 94
insecticide, arsenic as, 115
intelligence tests, study of Kellogg children's, 176n14
International Lead Zinc Research Organization (ILZRO), 57; and Bunker Hill disaster, 75, 89; and Bunker Hill Mining Co., 71, 90, 201n96; El Paso research funded by, 66; and follow-up study, 67; ghostwriting services of, 69; health effects of airborne lead denied by, 103–104; planned follow-up study of, 90–91; public relations efforts of, 58–59, 68; research backed by, 58; scientific studies sponsored by, 7
International Pacific Salmon Fisheries Commission, 23
invisible injury theory, 19, 180n40
IRATE (Island Residents against Toxic Emissions), 134

Jackson, Sen. Henry M., 47, 48
jobs: and environmental regulation, 113; *vs.* environmental standards, 132; *vs.* regulation, 148, 149; in Tacoma, WA, 113
jobs argument: health needs in, 165; industry's reliance on, 172
Johnson, Carl, 130

Kabwe, Zambia, lead smelter in, 168
Kehoe, Robert, 76
Kellogg, ID, 2, 73; ambient lead levels in, 83; childhood lead poisoning in, 74; heavy metal exposure in, 107; lack of organized opposition in, 97; lead intoxication in, 87. *See also* Bunker Hill smelter; Silver Valley
Kellogg Evening News, 96
Kennecott smelter, in Salt Lake City, UT, 166
Kern Place neighborhood, El Paso, 63; and MS incidence, 164
Kern Place nursery school students, testing of, 63
Kettering Laboratory, at Univ. of Cincinnati, 91

Lancet, 70
Landrigan, Philip, 61, 66, 71, 87, 89, 95, 211n49; on cancer threshold, 151; career of, 163; El Paso investigation of, 61–63; follow-up study of, 66, 67; and McNeil/ILZRO/ASARCO study, 69; NCV study of, 98; and Shoshone Project, 91, 99, 100; testimony at 1981 trial, 99; as threat to industry, 70
Lane, Wallace, 121
La Oroya, Peru: BLLs in, 169; environmental concerns in, 9

law, environmental, 143–144
lead: from El Paso's smelter emissions, 70; regulation of, 8; in smelter emissions, 122; standards on, 7; urine testing for, 75. *See also* air pollutants
lead, airborne: in American cities, 65; causes of, 65; exposure to, 58
lead absorption, undue, 102
lead arsenate, 24, 59
lead emissions: compared with arsenic, 138; in El Paso, 59; near Bunker Hill smelter, 83–84; regulation of, 69, 71
lead exposure: blame for, 109; of children, 2, 59, 60, 81, 84, 87; in El Paso, TX, 61; of infants, 94; life-long health effects of, 84; as lifelong health hazard, 172; long-term effects of, 164; near Bunker Hill smelter, 87; near nonferrous smelters, 8–9; in Ruston, 130; in Silver Valley, 80–81; subclinical, 67; symptoms associated with, 84, 99; from Tacoma smelter, 111, 211n49; testing for, 75; on Vashon Island, 130
Lead Industries Association (LIA): and Bunker Hill crisis, 89; federal standards challenged by, 105; research backed by, 58; scientific studies sponsored by, 7
lead industry: abuses within, 106; biased arguments used by, 63; and El Paso incident, 58; follow-up study controlled by, 67; harm to lead exposed children denied by, 104; indifference of, 174; research controlled by, 59; research funded by, 91
lead intoxication, in Kellogg, ID, 87
lead lines, seen on X-ray, 84
lead paint: and childhood poisoning, 59; market for, 59
lead poisoning: blame for, 109; children at risk for, 74; community ambivalence about, 107; effects in children, 66; in El Paso, TX, 4; epidemiologic studies of, 66; in Kellogg, ID; long-term effects of, 84, 88; near Bunker Hill smelter, 87; science of pediatric, 163; significance of subclinical, 163; in Silver Valley, 80–81; symptoms associated with, 63, 64, 86; testing for, 75
lead poisoning crisis: in El Paso, 111; role of press in, 108–109
lead poisoning epidemic: in El Paso, 54; in Kellogg, ID, 87; Silver Valley's, 74, 95
lead prices, and pollution control, 83
lead screening, in New York City, 60
lead smelting, decline of, 157
League of Women Voters, 53
learning difficulties, in lead-poisoned children, 103
Legge, Thomas, 115
Lerner, Sydney, 91
LIA. *See* Lead Industries Association
life-sustaining systems, 173
livestock: and arsenic exposure, 28, 130, 213–214n5; near Bunker Hill smelter, 85; poisoning of, 34, 50; and smelter pollution, 16

Lobe, Gene, 147
London fog, of 1952, 35
Loomis, Ted, 92, 132
Los Angeles, CA, smog in, 39
low birth weight, and lead exposure, 164
lung cancer: among miners, 116; and arsenic exposure, 116, 118, 132; in Ruston, WA, 13; in smelter workers, 123
lung problems, and urban air pollution, 36–37
Lyman, Donald R., 90

MacMillan, Donald, 16
Malone, Michael, 16, 157
Manufacturing Chemists Association (MCA), 38
manufacturing doubt: strategy of, 7, 74; success of, 172
Marcosson, Isaac F., 181n44
mass poisoning, with arsenic, 24
Mather, John, 103
maximum achievable control technology (MACT), 167
McClure, Sen. James, 107
McCormick, Marshall, 53
McDaniel, David, 88
McNeil, James L., 62, 66, 67, 92, 104; El Paso study of, 69, 89, 98; relationship with ASARCO, 68
McNeil/ILZRO/ASARCO study, 6, 66, 67–68, 69, 89, 98, 176n14
mean hair lead levels, 81
media: air pollution in, 35; blame for childhood lead poisoning in, 109; and blood test results in children, 88; response to health recommendations of, 108–109; smelter pollution in, 42; support for mining and smelting industry in, 96; Tacoma emissions debate in, 151. *See also* press; *specific newspapers*
medical records, access to, 125, 132
medicine, use of arsenic in, 114–115, 116
Mercier, Laurie, 106
mercury: health benefits of reducing, 168; in smelter emissions, 122
metals, volatilized, 15. *See also* heavy metals
metals industry, western nonferrous, 11. *See also* lead industry
Meuse Valley, Belgium, air pollution deaths at, 35, 36, 48
Mexican children, BLLs of, 70
Michaels, David, 7
middle class: concerned residents among, 36; in Tacoma, WA, 40
Milham, Samuel, 121, 122, 122f, 123, 130, 132, 133, 134, 135, 136; Ruston elementary school study of, 140; testimony on arsenic standard, 152, 211n49
mill tailings, dumping of, 79
mine owners, 79
miners, lung cancer among, 116
mine tailings, in Silver Valley, 160. *See also* tailings

mine waste, 81
mining, in Silver Valley, ID, 78, 78f, 80
mining and smelting industry: community support for, 96; decline of, 3; economic power of, 16; in Idaho's Silver Valley, 74, 80. *See also* smelting industry
Mining Association Convention (1974), 98
mining sites, pollution of, 169
monitoring studies, 127
Monsanto, 92; cancer research funding of, 209n26
Morgan, Murray, 32
mortality rate, and urban air pollution, 36–37
Moselle Valley, Germany, 116
mothers, blame for contamination placed on, 88. *See also* parents
Mount Isa, Australia, environmental concerns in, 9
Muir, Warren R., 69
Muir Commission, 69, 103–104
Murray Smelter, 19

National Academy of Sciences (NAS): on arsenic as environmental pollutant, 133; on lead concentrations, 81; 1972 report on airborne lead, 61; report on environmental arsenic of, 138–139
National Agricultural Chemicals Association, 117
National Cancer Institute (NCI), 118
National Emissions Standard for Hazardous Air Pollutants (NESHAP), for arsenic, 147–148
National Institute for Occupational Safety and Health (NIOSH), 55; arsenic concerns of, 131
National Lead Co., 58
National Public Radio (NPR), 167
National Resources Defense Council, 52
national smelter study, EPA's, 138
Nelson, Kenneth, 52, 123
nerve conduction velocity (NCV), and blood lead levels, 98, 99
nervous system, effects of lead on children's, 88
neurological effects, Bunker Hill–funded study of lead poisoned children, 103
neurological testing: funded by Bunker Hill Co., 101; in Idaho, 101
neuropsychiatric symptoms, associated with lead exposure, 164
New York State, EPA lawsuit of, 147
New York Times, 149
Nicola, Bud, 152
Nixon, Pres. Richard, 55
Nixon administration, 80, 92
nonferrous category, 14
Norilsk, Russia, metals smelting complex in, 168
North End Improvement Club, 153
Northern Pacific Railroad, 31
North Tacoma, contamination of soils in, 130

nuisance laws, 144, 145

occupational cohort study, 116
Occupational Safety and Health Administration (OSHA), 55; arsenic standard of, 132, 140; Bunker Hill citations of, 82; enforcement of standards by, 107; lead standard of, 106
ocean dumping, of arsenic, 23
odor complaints, 32
Office of Management and Budget (OMB), 173
Olson, L. V., 27, 40, 43, 49, 52
opinion poll, for Bunker Hill Co., 96
organic brain dysfunction, in lead-exposed children, 205n174
OSHA. *See* Occupational Safety and Health Administration
overregulation: allegations of federal government, 157; industry's claims of, 172

Panhandle Health District, ID, 85, 87
Panke, Ronald K., 75, 76, 87–88, 91, 92, 94, 98, 103, 104
parents: ambivalence of, 107; and arsenic exposure, 112; ASARCO's communication with, 135; attitudes toward testing, 100; blame for contamination placed on, 88, 136, 162; and lead poisoning crisis in Silver Valley, 108; and urinalysis screening, 76–77; Vashon Island compared with Ruston, 135
Paris, John Ayrton, 116
Patterson, Claire, 52
People magazine, 88
Peru, sulfur dioxide pollution from smelters in, 168
pesticides: arsenic, 24, 26, 119; phasing out of, 24
pets: arsenic poisoning of, 21, 210n37; in arsenic showers, 46; poisoning of, 50. *See also* livestock
Philippines, Lepanto ore from, 156
physicians: and arsenic exposure, 112; parents' dependence on, 109, 207n213. *See also* Doctor's Clinic
Pierce County Commissioners, petition presented to, 46–47
Pierce County Medical Society, 53
Pinto, Sherman, 21, 53, 117, 123, 137
Point Defiance Park, Tacoma, 129
Poisoned Places (documentary), 167
political discourse, and environmental health, 173
political power, of polluters, 168
politics: and environmental awareness, 56–57; and mining and smelting industry, 16
pollution, chronic: cleaning up of, 158; efforts to reduce, 168; tacit acceptance of, 165
pollution, smelter, dispersion of, 14. *See also* industrial pollution; smelter emissions

INDEX

pollution control: and Bureau of Mines, 14; dispersing sulfur dioxide for, 21; economics of, 48, 50; industry concerns about costs of, 168; public demands for, 82; resistance to investing in, 3; and role of industry, 32; in Tacoma, WA, 26; West vs. East Coast, 21. *See also* Puget Sound Air Pollution Control Agency

pollution dispersing strategy, through tall stacks, 4

pregnant women, bone lead in, 84

preschoolers, urinary arsenic monitoring among, 136. *See also* children

press: ASARCO supported by, 124; blame for childhood lead poisoning in, 109; and blood test results in children, 88; response to health recommendations of, 108–109; support for mining and smelting industry of, 96. *See also* media; *specific newspapers*

PSAPCA. *See* Puget Sound Air Pollution Control Agency

public, doubt manufactured for, 7, 74

public health, planning for, 173

public health history, smelting industry in, 6, 84

Public Health Service, U.S. (USPHS): on arsenic exposure, 210n37; Tacoma investigated by, 50–51; urinary arsenic levels study of, 119; Wenatchee study, 119

public interest, vs. industrial pollution, 173

public opinion, in Tacoma, WA, 109

public relations: ASARCO, 38, 123–124, 151, 159, 177n18; Bunker Hill Mining Co., 95–96, 99; industry's use of, 172; smelting industry, 7

Puget Sound: pollution of, 141–142; slag dumped into, 142

Puget Sound Air Pollution Control Agency (PSAPCA), 54; and arsenic standards, 123, 133; and ASARCO's compliance, 137; and EPA, 147; EPA settlement agreement with, 142; and EPA standards, 124; focus of, 112; formation of, 111; hearings of, 137; monitoring studies initiated by, 127; reprieve granted ASARCO by, 136; Tacoma smelter deadline set by, 52

Puget Sound region, 11, 12f; and arsenic NESHAP, 147; arsenic soil contamination in, 161f; environmental health investigations in, 125; research with bees in, 143; soil contamination in, 159–160

Puyallup Indian tribe, 141–142

pyrometallurgical process, 15

quality of life, impact of smelters on, 34

Radtke, Schrade F., 90

railroads, pollution associated with, 31

ranchers: disputes with smelting companies of, 17; pollution complaints of, 78–79

rat poison, arsenic as, 24

Reagan, Pres. Ronald, 139

Reagan administration, 146; EPA during, 140

Reed, David, 146

Reed, Sam, 211n49

Reeds, Thomas, 84, 85

refining, in copper production, 15

regulation: Bunker Hill opposition to, 96; cost of, 149; effectiveness of, 167; and lost-jobs argument, 148, 149; and smelter closure, 155; in Tacoma, 152–153. *See also* Environmental Protection Agency

Reitan, Ralph, 99

Reitan testing, 99

renal disease, and lead exposure, 164

research: on arsenic, 118–119; environmental, 113; on geographic extent of smelter pollution, 143; on health effects of air pollution, 168; industry-funded, 91–92; industry-sponsored, 89. *See also* American Smelting and Refining Co. research

research, independent: ASARCO's resistance to, 138; credibility of, 172; funding for, 171; government investment in, 54; inconclusive results for, 140–141

respiratory cancer, in Tacoma smelter workers, 123

respiratory symptoms, in Tacoma children, 41. *See also* lung cancer

Richards, Bill, 107

risk assessment, cost-benefit analysis in, 149–150

Robinson, Philip E., 68

Rochlin, Kevin, 155

Rockefeller, William, 3

Rockefeller family, 3

Rodgers, William, 121

Rogers, Henry, 3

Rosenblum, Bernard, 61, 62, 67

Rossellini, Gov. Albert D., 47, 48

Roush, George, 92

Rowlands, City Manager David, 43, 44, 45, 50, 53

Ruckelshaus, William, 80, 139, 147–148, 149–150

Russell, Robie, 158

Russia, sulfur dioxide pollution from smelters in, 168

Ruston, WA, 1, 11, 26, 129; arsenic concentrations in air in, 157; arsenic exposure in, 5; children's exposure to arsenic in, 133; contamination of soils in, 130; regional contamination problem identified in, 125–126; Superfund cleanup in, 160; as Superfund site, 158, 159; urinary arsenic in children of, 120–121, 122, 123

Ruston Elementary School, 121

Ruston residents, 11, 13; ambivalence about cleanup of, 159

Sachs, Henrietta, 67, 92, 104

Salt Lake City, UT, Kennecott smelter in, 166

Salt Lake Valley Smelter Commission, 18

science: and ecological thinking, 51–52; environmental, 126; on environmental

science (*continued*)
  arsenic, 145–146; as excuse for inaction, 171; industry-sponsored, 14; of pediatric lead poisoning, 163; in public health debates, 6; role in struggles over pollution, 18. *See also* research
scientists: ASARCO, 125; industry-affiliated, 60
screening: in Kellogg, ID, 74, 75, 196n3. *See also* urinalysis screening
scrotal cancer, 116
"sea captain" theory of smoke control, 19
Seattle–King County Health Dept., 129, 130; on contaminated soils, 146
Seattle Post-Intelligencer, 149
section 112, of Clean Air Act, 149
Selby smelter, 20, 22, 27
Shinkoskey, Robert E., 40, 44–45, 49, 52
Shoshone County, 79
Shoshone Project, 91, 99; blood lead levels in, 93, 100; community response to, 97–98; director appointed for, 92; disabled child omitted from, 94; findings of, 98, 99, 100, 102; flaws in, 95; funding of, 6, 92, 176n14; ILZRO's involvement in, 92, 201n102; interim report for, 94; investigation of, 107; memorandum of understanding for, 92–93; news releases coming from, 109; omissions in, 93–94; stated aims of, 93; technical committee of, 91–92
Sierra Club, 52, 134
Silent Spring (Carson), 24, 51
Silver King Elementary School: average air lead level at, 103; fear of parents of children at, 108; urinalysis screening at, 76–77
Silver King School Board, 94–95
Silver Valley, Idaho's, 73, 77, 78f; blood lead levels in, 89; cleanup of, 164; dependence on mining in, 79; lead poisoning epidemic in, 74; lead poisoning of children in, 90; twenty-year follow-up study of children of, 163. *See also* Kellogg, ID
Silver Valley cleanup: and ASARCO bankruptcy settlement, 160; and EPA, 160, 162
Silver Valley Lead Study, 105
Silver Valley residents, long-term effect of lead exposure for, 164
skin diseases: arsenical dermatoses, 123, 210n37; arsenic associated with, 22–23; arsenic poisoning, 114–115; cancer, 114–116
slag: defined, 15; dumping of, 142
smelter emissions: and ASARCO's research, 18; follow-up of U.S. communities exposed to, 163; heavy-metal, 61–62; and ILZRO, 57; nonferrous, 61; in Tacoma, WA, 22. *See also specific smelters*
smelter industry: on airborne lead, 60; and EPA regulations, 165; and lead in gasoline, 60; pushback of, 146; reduction in pollution by, 166; science sponsored by, 14
smelters: arsenic standard for, 139; community health effects from, 71, 163; emissions from, 56; nonferrous, 14, 56; pollution caused by, 4; "tidewater," 13
smelters, nonferrous: arsenical air pollution caused by, 147; EPA settlement with, 165; forces driving closure of, 155–156; impact of global changes on, 155
Smeltertown, 62–63, 64; and MS incidence, 164; razing of, 66
Smeltertown children: BLLs in, 62, 63–64, 177n18; clinical follow-up for, 67, 68–69; J. McNeil's view of, 86
Smeltertown residents, response to crisis of, 65–66
The Smelter Worker (newsletter), 132
smelting: decline of, 157; pollution damage from, 16, 17; public health history of, 5, 6; at Tacoma, WA, 15; toxic metal pollution from, 15–16; in U.S., 16. *See also* mining and smelting industry
smelting and refining industry: environmental regulation of, 146; slow decline of, 146–147
smelting industry, 3; and cleanup costs, 158–159; early pollution control in, 13; and El Paso disaster, 57; and EPA lead standard, 105; global, 168; "manufacturing doubt" strategy of, 7; overseas, 9; precipitators adapted for, 20; in public health history, 84; and public health issues, 56; public relations of, 7, 123–124; science sponsored by, 6, 7; "success" of, 172; sulfur dioxide control attempts of, 180n35; and sulfur dioxide controls, 56
smelting sites, pollution of, 169. *See also specific sites*
smog, 39, 42
smoke, from sulfur dioxide, 154
smoke abatement efforts, 33
"smoke farmers," 17, 25
smoke ordinance, in Tacoma, WA, 37, 38
smoke problems, in Tacoma, 33
smokestacks, smelter, 2, 113; in El Paso, 57–58; over Tacoma, 1–2; toxic metals dispersed by, 17. *See also* tall stacks
Society for Occupational and Environmental Health, 69
soils: smelter contamination of, 130, 145; Washington State standard for arsenic in, 223n26
soils, contaminated: with arsenic, 223n26; in El Paso, TX, 162; monitoring storage of, 163; in Puget Sound Region, 161f; in Ruston, WA, 130; Seattle–King County Health Dept. on, 146; in Silver Valley cleanup, 162; from Tacoma smelter, 159; on Vashon Island, 213–214n5
Soil Safety Program, Washington State, 161
soil theory, ASARCO's, 64–66
standard of living, public's expectations for, 34–35
State Air Pollution Control Board, Washington's, 45, 48
steelworkers union, and Tacoma smelter closure, 156

INDEX 237

St. Joseph Lead Co., 58, 105
Stokes, Lee, 95
Strauss, Simon D., 124
St. Regis Pulp and Paper Co., 32
Strong, Terrance R., 211n49
sudden infant death syndrome, and sulfur dioxide air pollution, 42
sulfur dioxide: in copper smelting, 15; "invisible injury" caused by, 19; regulation of, 8; Tacoma ordinance for, 54; from Tacoma smelter, 142. *See also* air pollutants
sulfur dioxide capture: in Tacoma, WA, 37; technology, 18
sulfur dioxide control: ASARCO's position on, 18, 180n35; attempts at, 180n35; baghouses, 19; costs of, 180n36; resistance to, 82; "sea-captain" theory of, 19; in Tacoma, WA, 25
sulfur dioxide emissions, 3; anthropogenic, 169; ARCO's data on, 47; from Bunker Hill smelter, 157; in El Paso, TX, 4, 56; environmental effects of, 16; from smelters, 3, 19; sources of, 55–56; from Tacoma smelter, 39–40; and toxic metals control, 20
sulfur dioxide exposure: respiratory symptoms associated with, 41; symptoms associated with, 48; in Tacoma, WA, 42
sulfur dioxide recovery, in Tacoma, WA, 25
sulfur dioxide regulations, ASARCO's compliance with, 137
sulfur fumes, experienced as taste, 41
Sunshine mine, 86
Superfund cleanups, 13; in Ruston, 160; in Silver Valley, 160
Superfund program, 158
Superfund site: Ruston as, 159; in Tacoma, WA, 12f
"superstack," proposal for, 21, 181n
Supreme Court, U.S., and smelter closure, 26–27
surgeon general, U.S., lead poisoning guidelines of, 93
Suskind, Raymond, 91
sustainability, planning for, 173
Sutherland, Mayor Doug, 151
Swain, Robert E., 17, 18, 19

Tacoma, WA, 2, 11, 129; air pollution in, 27–28; arsenical dermatoses in, 210n37; arsenic control in, 25; arsenic exposure in, 5, 112, 126, 131; environmental contamination problem in, 124; environmental regulation in, 152–153; industrial plants in, 31; location of, 31; pollution controls at, 26; regional contamination problem identified in, 125–126; slag produced in, 15; smoke ordinance in, 37, 38; sulfur dioxide ordinance of, 25, 54; on Superfund list, 158; unemployment in, 148; urinary arsenic in children of, 120–121, 122, 123; West End development of, 40–41, 42, 46
"Tacoma Air Pollution Study" (1970), 212n66
Tacoma City Council, petition presented to, 46
Tacoma city government, 45
Tacoma Daily Ledger, 25, 33
Tacoma Engineers Club, 36
Tacoma News Tribune, 28, 34, 39, 45, 61, 124, 148
Tacoma–Pierce County Health Dept., 143, 144, 152
Tacoma Public Library, 43
Tacoma residents: cancer risk from arsenic for, 148, 151; community organizing of, 36, 53; complaints of, 28; demographic changes, 35–36; focus on industrial air pollution of, 39; new awareness of, 52
Tacoma smelter, 11, 12f, 33; and arsenic exposure, 133, 134; arsenic pollution from, 21; and cancer risk; capacity of, 26; closure of, 155; compared with Bunker Hill smelter, 81–82; costs of, 150; damage claims against, 22; decision to close, 152; and elevated BLLs in children, 122; environmental impacts of, 141; environmental impact statement for, 145; environmental specialist of, 135, 136; lawsuits against, 26; location of, 11, 12f, 13; niche for, 146–147; on-site containment facility of, 160; precipitator installed at, 20; profitability of, 21–22; property damage from, 27; public clamor for pollution control of, 25; public relations campaign of, 124; smokestack fire of, 144; smokestack of, 17; soil contamination footprint of, 159–160; source of ore for, 147; and summer of 1966, 52; and surrounding community, 49f. *See also* workers, Tacoma smelter
Tacoma smelter emissions: arsenic, 130; arsenic dust, 144; EPA estimates for, 126; health risks of, 150; heavy metal, 111–112; impact on area vegetation of, 40, 50, 126; impact on Vashon Island of, 130; independent research on environmental effects of, 113; reductions in, 133–134; soil contamination from, 4; struggle to control, 113–114; sulfur dioxide, 39, 112, 142
tailings: dumping of, 79; monitoring storage of, 163; in Silver Valley, 160
tall stacks, 3; at Bunker Hill smelter, 82; at El Paso smelter, 57–58; emissions dispersed from, 14; impact of, 3–4; industry's reliance on, 21; limitations of, 25; rationale for, 181n44; and smoke conditions, 181n51; and sulfur dioxide pollution, 19, 20; symbolism of, 58; and urinary arsenic concentrations, 121; during war, 33
temperature inversions, and air pollution, 81, 198n45
Terry, Surgeon Gen. Luther, 50
testing for lead exposure: sample for, 76; test performance in children, 205n174
Texas, ASARCO consent decree with, 65, 192n45
Texas Air Control Board, 65
threshold concept, 137; ASARCO's emphasis on, 151; ASARCO's use of, 150

timber industry, 31–32
Tittman, Edward, 44
tobacco industry, indifference of, 174
Tobin, Bill, 145
toxic metal emissions: alveolar deposition of, 126; in Silver Valley, 83; from smelters, 3. *See also* heavy metals
toxic tolerance program, 150
toxic waste, communities devastated by, 165
trade groups, industry, scientific studies sponsored by, 7
Trail Smelter, British Columbia, 180n36
Travis, L. C., 124
trespass: intentional, 145; legal concept of, 145; toxic, 146
Tuberculosis and Respiratory Diseases Association, 134
Tyler, Richard, 28

"Uncle Bunker," 79, 80
unemployment, in Tacoma, 148. *See also* jobs; workers
union, Bunker Hill, 107
Union Bag & Paper Co., 33
United Steelworkers of America (USW), 106
universities, cooperation with industry of, 38
university researchers, ASARCO's support of, 18. *See also* research
urinalysis: of arsenic concentrations, 210n34; for lead, 75
urinalysis screening: for arsenic detection, 118; conducted by ASARCO, 134–136, 216n31; for lead levels, at Silver King Elementary School, 76–77; "normal" arsenic in, 119, 120, 122; on Vashon Island, 213–214n5
urinary arsenic, "normal," 122

vacuum cleaner dust, 121
Vallee, Bert L., 117
Vashon Island, 129, 155; attraction of, 129–130; contamination of soils on, 130, 213–214n5; environmentalist ethic on, 131; parents on, 135; residents' concerns on, 143
Vashon Island Community Council, 150
vegetation damage, 81; from smelter emissions, 144; in Tacoma, 40, 50, 126
victory gardeners, 34
von Lindern, Ian, 83–84, 85, 95, 105, 109, 163

Walter, Stephen D., 105
Washington Environmental Council, 134
Washington Post, 80, 107
Washington State: early appeals to, 47; and industrial air pollution, 39; and pollution control, 37–38; unemployment in, 148
Washington State Health Dept.: monitoring studies initiated by, 127; downplays health risks of smelter emissions, 144–145; and smelter's health risks, 141
Washington State Pollution Control Commission, 23, 27
waste, from copper smelting, 15
waste disposal, industrial, 32
Watergate bugging incident, 80
water pollution control, public demands for, 82
weather: and air pollution, 81, 198n45; and sulfur dioxide control, 19
Wegner, Glen E., 92, 94, 95, 97, 99, 107
Wenatchee study, 119, 122
Weyerhaeuser, 38
Whelan, Paul, 109
White House Council on Environmental Quality, 69
Williams, Mayor Bert, 64
Wingard, Jean, 144
Woodruff, Frank, 73
workers, Anaconda smelter: arsenic-exposed, 123; NCI's 1969 study of, 137
workers, Bunker Hill: ambivalent position of, 106; blood lead levels of, 106; lead exposure of, 82, 83; pressure on, 106
workers, pesticide production, study of arsenic in, 131–132
workers, smelter: arsenical skin cancer in, 115; arsenic-exposed, 118; lung cancer in, 123; NCI study of U.S., 118; study of arsenic in, 131–132
workers, Tacoma smelter: access to medical records of, 125; arsenic levels of, 120; ASARCO-funded study of, 117; cancer risk for, 137; and environmentalists, 156; lung cancer among, 132; respiratory cancers among, 123
worker safety regulations, Bunker Hill opposition to, 96
World Health Organization (WHO), 77
World War I, and copper industry, 26
World War II, and manufacturing jobs, 33

Yankel, Anthony, 105
Yoss case, 109

ABOUT THE AUTHOR

Marianne Sullivan is an assistant professor of public health at William Paterson University of New Jersey. Prior to that she was on the faculty of Hofstra University and worked as an epidemiologist for Public Health—Seattle & King County.

Available titles in the Critical Issues
in Health and Medicine series:

Emily K. Abel, *Suffering in the Land of Sunshine: A Los Angeles Illness Narrative*

Emily K. Abel, *Tuberculosis and the Politics of Exclusion: A History of Public Health and Migration to Los Angeles*

Marilyn Aguirre-Molina, Luisa N. Borrell, and William Vega, eds. *Health Issues in Latino Males: A Social and Structural Approach*

Susan M. Chambré, *Fighting for Our Lives: New York's AIDS Community and the Politics of Disease*

James Colgrove, Gerald Markowitz, and David Rosner, eds., *The Contested Boundaries of American Public Health*

Cynthia A. Connolly, *Saving Sickly Children: The Tuberculosis Preventorium in American Life, 1909–1970*

Tasha N. Dubriwny, *The Vulnerable Empowered Woman: Feminism, Postfeminism, and Women's Health*

Edward J. Eckenfels, *Doctors Serving People: Restoring Humanism to Medicine through Student Community Service*

Julie Fairman, *Making Room in the Clinic: Nurse Practitioners and the Evolution of Modern Health Care*

Jill A. Fisher, *Medical Research for Hire: The Political Economy of Pharmaceutical Clinical Trials*

Alyshia Gálvez, *Patient Citizens, Immigrant Mothers: Mexican Women, Public Prenatal Care, and the Birth Weight Paradox*

Gerald N. Grob and Howard H. Goldman, *The Dilemma of Federal Mental Health Policy: Radical Reform or Incremental Change?*

Gerald N. Grob and Allan V. Horwitz, *Diagnosis, Therapy, and Evidence: Conundrums in Modern American Medicine*

Rachel Grob, *Testing Baby: The Transformation of Newborn Screening, Parenting, and Policymaking*

Mark A. Hall and Sara Rosenbaum, eds., *The Health Care "Safety Net" in a Post-Reform World*

Laura D. Hirshbein, *American Melancholy: Constructions of Depression in the Twentieth Century*

Timothy Hoff, *Practice under Pressure: Primary Care Physicians and Their Medicine in the Twenty-first Century*

Beatrix Hoffman, Nancy Tomes, Rachel N. Grob, and Mark Schlesinger, eds., *Patients as Policy Actors*

Ruth Horowitz, *Deciding the Public Interest: Medical Licensing and Discipline*

Rebecca M. Kluchin, *Fit to Be Tied: Sterilization and Reproductive Rights in America, 1950–1980*

Jennifer Lisa Koslow, *Cultivating Health: Los Angeles Women and Public Health Reform*

Bonnie Lefkowitz, *Community Health Centers: A Movement and the People Who Made It Happen*

Ellen Leopold, *Under the Radar: Cancer and the Cold War*

Barbara L. Ley, *From Pink to Green: Disease Prevention and the Environmental Breast Cancer Movement*

Sonja Mackenzie, *Structural Intimacies: Sexual Stories in the Black AIDS Epidemic*

David Mechanic, *The Truth about Health Care: Why Reform Is Not Working in America*

Alyssa Picard, *Making the American Mouth: Dentists and Public Health in the Twentieth Century*

Heather Munro Prescott, *The Morning After: A History of Emergency Contraception in the United States*

David G. Schuster, *Neurasthenic Nation: America's Search for Health, Happiness, and Comfort, 1869–1920*

Karen Seccombe and Kim A. Hoffman, *Just Don't Get Sick: Access to Health Care in the Aftermath of Welfare Reform*

Leo B. Slater, *War and Disease: Biomedical Research on Malaria in the Twentieth Century*

Paige Hall Smith, Bernice L. Hausman, and Miriam Labbok, *Beyond Health, Beyond Choice: Breastfeeding Constraints and Realities*

Matthew Smith, *An Alternative History of Hyperactivity: Food Additives and the Feingold Diet*

Rosemary A. Stevens, Charles E. Rosenberg, and Lawton R. Burns, eds., *History and Health Policy in the United States: Putting the Past Back In*

Barbra Mann Wall, *American Catholic Hospitals: A Century of Changing Markets and Missions*

Frances Ward, *The Door of Last Resort: Memoirs of a Nurse Practitioner*

CPSIA information can be obtained at www.ICGtesting.com
Printed in the USA
BVOW02s1334171213

339031BV00001B/1/P